FIRE IN THE NIGHT

Stephen McGinty is an award-winning journalist and the author of *This Turbulent Priest: A Biography of Cardinal Thomas Winning*, *Churchill's Cigar* and *Camp Z: How MI5 Cracked Hitler's Deputy*. He writes for the *Sunday Times Scotland* and produces television documentaries including the BAFTA Scotland-winning *Dunblane: Our Story*. *Fire in the Night*, the feature-length documentary based on his book, won the 'Audience Award' at the Edinburgh International Film Festival.

Also by Stephen McGinty

This Turbulent Priest

Churchill's Cigar

Camp Z

STEPHEN McGINTY

FIRE IN THE NIGHT

THE PIPER ALPHA DISASTER

First published 2008 by Macmillan

First published in paperback 2009 by Pan Books

This updated edition published 2018 by Pan Books
an imprint of Pan Macmillan
20 New Wharf Road, London N1 9RR
Associated companies throughout the world
www.panmacmillan.com

ISBN 978-1-5098-6822-3

Picture Acknowledgements
Paul Berriff / Scottish Television: p.6 (both)
The Cullen Inquiry: p.1; p.4 (both); p.5 (both); p.7 (top)
Rex Features: p.8 (top)
Scotsman: p.7 (bottom)
Sue Jane Taylor: p.2 (both); p.3 (both); p.8 (bottom)

9 8 7 6 5 4 3 2 1

A CIP catalogue record for this book is available from the British Library.

Typeset by SetSystems Ltd, Saffron Walden, Essex
Printed and bound by CPI Group (UK) Ltd, Croydon, CRO 4YY

Visit **www.panmacmillan.com** to read more about all our books
and to buy them. You will also find features, author interviews and
news of any author events, and you can sign up for e-newsletters
so that you're always first to hear about our new releases.

For the 167

Burning in Water, Drowning in Flame

Charles Bukowski

Contents

Note on Sources

This is an incomplete book. The story of the two hours between 10 pm and midnight on 6 July 1988, during which time 167 men lost their lives can never be told in its totality. There are just too many silent voices. While the effects of heat on steel can be replicated under laboratory conditions and the results extrapolated, the effect of heat and fear on a man's character cannot be known without his testimony. We will never know the true story of each man's final hour. The best we can do is assemble those pieces like a jigsaw, in order to reflect on the black holes that lie between.

The narrative of *Fire in the Night* was constructed using a number of different sources; however, the principal source of information was the two-volume report written by Lord Cullen and published in November 1990 as *The Public Inquiry Into The Piper Alpha Disaster*. The chronological events of 6 July 1988 and the involvement of different individuals was taken from their own detailed testimony to the inquiry, the daily verbatim records of which are available at the Scottish Archives in Charlotte Square in Edinburgh. Additional information was drawn from contemporary newspaper reports by the *Scotsman*, the *Herald*, the *Aberdeen Press & Journal*, the *Aberdeen Evening Express*, *The Times*, *The Sunday Times* and the *Independent*, as well as television documentaries produced by the BBC, Scottish and Grampian Television. A number of books were of crucial assistance including *Piper Alpha:*

A Survivor's Story by Edward Punchard (Star Books, 1989); *Off-shore: A North Sea Journey* by A. Alverez (Hodder & Stoughton, 1986); *Fool's Gold: The Story of North Sea Oil* by Christopher Harvie (Penguin, 1994); *Paying For The Piper: Capital and Labour in Britain's Offshore Oil Industry* by Charles Woolfson, John Foster and Matthias Beck (Mansell Publishing, 1996); *The Dark Side of Power: The Real Armand Hammer* by Carl Blumay with Henry Edwards (Simon & Schuster, 1992); *Hammer: Witness to History* by Armand Hammer with Neil Lyndon (Simon & Schuster, 1987); *Oilwork: North Sea Diaries* by Sue Jane Taylor (Birlinn, 2005); *The Oilmen: The North Sea Tigers* by Bill Mackie (Birlinn, 2004); *Rescue: The True-Life Drama of Royal Air Force Search and Rescue* by Paul Beaver and Paul Berriff (Patrick Stephens Ltd, 1990); and *Red Adair: An American Hero* by Philip Singerman (Bloomsbury, 1989). The anatomy of the fire is drawn from: 'The explosion and fire on the Piper Alpha platform, 6 July, 1988: A Case Study' by D.D. Drysdale and R. Sylvester-Evans, published in the Royal Society of London in 1998. Further insight into the psychological effect of the disaster can be found in Professor David Alexander's paper: 'Psychiatric intervention after the Piper Alpha disaster', published in the *Journal of the Royal Society of Medicine* in 1991.

An invaluable and under-publicized treasure trove of information on the North Sea industry is the Lives in the Oil Industry Oral History Project organized by Terry Brotherstone, director, and Hugo Manson, the project manager for the University of Aberdeen who conducted detailed interviews with almost 200 individuals including the late Bob Ballantyne, Sue Jane Taylor and Kate Graham.

Where dialogue appears in direct quotes, it is the individual's recollection of the exact words spoken at the time or those of a person present. Where specific thoughts are attributed to an individual, it is on the basis that they have spoken about what they were thinking at that time.

PROLOGUE

It began on the stroke of 10 p.m. with a flash of white light and a bang that punched through the sea and sent a flotilla of sonic waves rippling down 50 feet to where a diver, Gareth Parry-Davies, was at work. While the sky above was a bruised blue and had not yet turned to black, delayed by the long light nights of high summer, Parry-Davies was resident in a darker world illuminated only by the torch attached to his yellow steel helmet. Dressed in a standard red and black diving suit, through which a network of pipes circulated hot water to shield against the sea's bitter cold, he was firing a grit gun at a horizontal metal strut that made up one small part of the massive steel legs which descended into the blackness where they were rooted into the seabed 400 feet below. It was now forty minutes since he had been lowered inside the sphere-shaped diving bell, slipped out through the floor hatch and swum the short distance to where the metal member – which was flooded and suspected of harbouring a crack – was positioned. The noise of his own steady breathing, amplified by the bowl of the helmet, was broken up by the flow of questions, checks and the odd joke made by John Barr, the diving supervisor. This was carried down from dive control by a communication line, part of the umbilical, a thick, rubber-covered rope of multi-strands that carried hot water, air and sound, and was attached to the diver's suit. Parry-Davies had completed about 40 per cent of the task when he was shaken

1

by a bright white light which, despite the helmet's lack of peripheral vision, he recognized as coming from the right.

The heavy bang was simultaneous. In his hand, the grit gun died. He began to breathe in the heavy, shallow breaths of the startled. The first suspicion was that the hose powering the grit gun had burst. He had begun to look round, catch his breath and assess the situation when twenty seconds after the first, a second bang and a larger curtain of light settled briefly over his helmet. On the intercom Parry-Davies heard Barr instruct him to ditch the tools (an indication of the severity of the situation given the cost of recovery from the seabed) abandon the job site and return immediately to the diving bell. Parry-Davies was already kicking his legs through the water, the tools tumbling into the gloom, when he received the order. He issued an instruction of his own because as he swam back to the bell he could feel by the weighty drag that the umbilical was not yet being recovered. He feared that it might snag on the various bolts and steel braces dotted around the submerged structure, but Barr said he would see to it.

The dive control room hung like a gondola underneath the main diving module, a suite of small cramped offices on the 64-foot level of the oil platform which was accessed via a ladder and an aperture in the main module floor. Barr had been knocked off his chair by the bang, which emptied shelves of their books and files. The control room was slung low so that its windows provided a clear view onto the dive skid, the metal-framed structure where divers prepared to exit and enter the sea, but now water from the sprinkler system was spraying down onto the windows and obscuring his view. However, the water would not flow for long.

Edward Amaira, a fellow diver who had been sheltering from the wind in the 'Wendy House', a nearby storage facility, was the first to reach the dive skid and begin to organize the retrieval of Parry-Davies. While he attempted to start up the bell winch, his colleague Joe Wells came to assist with the second winch that was used to draw up the umbilical, but Wells began to struggle. The electricity had cut out rendering the machinery inert, so the umbilical would have to be pulled up manually. Amaira lent a hand as Wells stood on the bottom railing of the handrail, reached up towards the hydraulic winch, pushed forward the lever which released the lock and allowed the wheel to spin freely. The pair then grabbed the umbilical and began to pull down. Below the surface Parry-Davies felt the drag drop, the umbilical move, and breathed a little easier.

The duo on the dive skid became a trio with the arrival of Keith Cunningham, already wearing an emergency breathing apparatus of face mask and oxygen tank. A veteran of a previous explosion four years earlier, he had been joking that they were due for a repeat just moments before it occurred. Although there was a strong smell of smoke the dive skid was clear of actual smoke for the moment, yet there was a hint of flame. The main oil line, a 30-inch steel pipe, ran horizontally above the skid. On cold days the divers would reach up and touch it to warm their hands. Oil was now running down the outside of its steel skin, while at the elbow joint – where the line turned from horizontal to vertical and then disappeared up into the floor of the next level – small orange flames darted. Catching sight of the flames served only to quicken Cunningham's hands as he activated the bell winch, which in the event of an electrical failure was powered by compressed air.

Inside the bell, Parry-Davies was in constant communication with John Barr, who told him that he was still in 'free time', meaning he had not been underwater long enough to require decompression. (While diving, the gas breathed into the lungs is pumped into the bloodstream. This happens quickly under the pressure of the weight of water, but it takes longer for the lungs

3

to filter the gas out of the bloodstream during an ascent. If the pressure is reduced too rapidly gas bubbles are trapped in the joints, cutting off the blood supply with agonizing, and potentially fatal, effect.)

When the bell was raised back onto the skid, Parry-Davies saw by the tense expression on his workmates' faces that something was wrong, but he was not sufficiently panicked to begin quizzing them. Instead, with their silent assistance he set about stripping off his heavy gear. As Amaira was unaware of exactly how long Parry-Davies had been under, and had not heard Barr's instructions, he followed standard procedure: the diver was to go immediately to the decompression chamber.

•

Stan MacLeod, the diving superintendent with responsibility for the entire dive operation, had been leaning against a filing cabinet in the office of Barry Barber, his opposite number from Occidental, when the explosion occurred. He often popped in for a chat and to check on the orange and apple shoots he was cultivating in a flowerpot positioned beneath an old UVA lamp he had found in Barber's office. Barber, meanwhile, was more interested in stock accumulation than any organic cultivation. Currency exchange was his special interest, and in quiet moments he would puff on a menthol cigarette and scan the pink pages of the *Financial Times*, one of the many papers delivered daily by Occidental. In the midst of a conversation between the two men and two other staff (Dick Common, Barry Barber's clerk, and Edward Punchard, an inspection controller) the room shook, the shelves collapsed, the lights tumbled from their fixtures and a number of the ceiling's metal panels clattered to the floor, their short drop broken by Punchard's head.

Once they had recovered from the shock, the men began to assess the situation and be ready, if necessary, for evacuation. As

Punchard set off to find a hard hat, MacLeod went looking for 'the fuckin' breathing apparatus'. Unable to find it, he left the office via the south exit, which took him to the decompression chambers where he saw small pieces of debris that resembled pipe lagging lying smouldering on the floor. The heavy steel door from one of the two decompression chambers lay there too, having been blown from its hinges. There was also the 'ominous glow of flames'.

After checking that Parry-Davies was being recovered, MacLeod returned to the main office to confer with Barry Barber. He noticed that in the intervening few minutes the sprinkler system had failed to activate beyond a slight trickle. There had been no tannoy announcements; the rig, always roaring with so much activity that large plastic ear-protectors were mandatory, was ominously quiet. There was a stillness behind which, depending on a person's proximity, could be detected the faint crackle of fire.

The standard procedure in the event of an incident was for Barry Barber to contact the radio room staff, who would then inform him of the cause of the disturbance and the safest route to a lifeboat. But it was not to turn out this way. When Barber eventually got through to the radio room, the person who answered sounded panicked, confused and unable to assist in providing a safe route. So the two men implemented their own safety system; the divers would leave the site by going along the office corridor, in through the south door and out of the north door, which would allow Barber and MacLeod to do a head count and so check that each man had been able to leave his post safely.

MacLeod then called John Barr to check on the status of Parry-Davies. Barr reported that the diver was out of the water, but he had just learned that he had been instructed in error to go into decompression. It appeared that Parry-Davies had been accompanied to the decompression chambers, which were up a flight of steel stairs, by Christopher Niven, another diver. The missing door on the first chamber forced him to use the second, whose internal lights no longer worked. He climbed in, the door was locked and, over the course of one minute, the pressure inside the chamber was blown down to the equivalent of 40 feet underwater. He sat down and then tried to relax and breathe steadily, but it was disconcerting to be locked in a steel prison in the midst of an obvious crisis.

Less than a minute later a fresh face appeared at the small window of reinforced glass and peered in. It was Stan MacLeod. He wore a face mask and breathing apparatus, and gave Parry-Davies the standard diving signal for 'OK' – the thumb and forefinger touching in a circle with the three remaining fingers raised. He then gave him the thumbs-up, to indicate that they were bringing him back up to atmospheric pressure.

Outside the chamber, Stan MacLeod and John Barr were both far from OK. Next to the chamber were the oxygen quads, a collection of twelve bottles (five feet six inches tall, ten inches in diameter) filled with oxygen under high pressure for use in the decompression chamber. A hose usually played over them to keep them cool, but there was no longer any water pressure. When he first arrived MacLeod had tried to cover them with a fire blanket, but burning oil had begun to drop from above on to the blanket that was now ablaze. While John Barr was at the controls of the decompression chamber, Stan MacLeod began to look around for a scaffold pole; his plan was to use it to break off the pipework attached to the chamber and so speed up the return to atmospheric pressure that would release the door and allow Parry-Davies to escape.

MacLeod was terrified. His heart was racing and his face became contorted with fear and rising panic as his search grew increasingly frantic. He was convinced that at any second the bottles, weakened by heat, would crack and that all three of them would be killed in the explosion. He did not believe they would survive.

Then suddenly the door clicked open, Parry-Davies squeezed out and the three men rushed off together.

It was 10.06 p.m.

•

Each diver who passed through the dive control centre and had his name ticked off Barry Barber's list bore witness to an escalating fire. As he had evacuated the dive skid and climbed the stairs to the 68ft-level, Edward Amaira was confronted by fifty drums of rig wash, a chemical substance designed for scrubbing oil off the structure, now merrily ablaze, with flames dancing three feet high above the drums. Keith Cunningham reported thick, dark clouds of smoke billowing out to the west of the rig, as well as burning debris dropping into the water. Alastair Mackay reported that the umbilicals coiled up on the dive skid were now alight. The 'Wendy House' would soon join the conflagration.

Like everyone else on the rig, the divers had a designated lifeboat to which they were expected to report. It was two levels up, a height of 39 feet above the dive module and situated at the north-west corner. Ed Punchard was the first to try to reach it; this entailed a walk along the cellar deck (the second lowest deck on the rig) to the corner, then up two flights of stairs, first to the 83ft-deck and then on up to the 107ft-level. En route he passed a foreman, his face smeared with soot and blood, and a rigger who was hopping while trying to support an injured leg. As Punchard reached the stairs he was stopped by three men in blue Occidental overalls coming down and carrying an injured man.

'Forget it. It's all blocked with smoke and fire,' said one. So Punchard turned round and began to head back to the diving offices. From the corner he could see in the distance a fractured electrical cable jittering and jumping on the deck as burning debris fell onto the divers' changing rooms.

When Punchard returned to the dive centre and asked if Barry Barber had been issued with a safe route by the radio room, he was told no: 'I don't know a thing.' The fire developing by the gas cylinders now prevented them from going south, so once again Punchard set off along the cellar deck to the north-west corner.

Barber said he would wait until all the team had come by. If Punchard had not returned he would assume the route was safe. He then tried to lighten the mood with a joke: 'When the rest of the lads are up we'll be hot on your heels.'

Punchard, accompanied by another diver, Andy Carroll, set off again. The wind was blowing from the south-west, which meant that at least the corner was clear of the thick smoke now obscuring the centre of the platform. At the stairwell a group of men had gathered. While several appeared shocked, the majority proved to be calm. An injured man was receiving treatment from two colleagues; a few were looking down over the waist-high steel barrier at the grey sea 68 feet below. When one man said he didn't think the situation could be that serious as there were no fire alarms, a colleague took a pin to his optimism and suggested that perhaps they didn't work.

Meanwhile, Punchard was becoming concerned about an air-compressor which was racing away a few feet from where they were all standing. In the past he had seen it spit out sparks from its exhaust, and now he was worried in case it ignited a gas leak. As he began to search for an 'off' switch Davey Elliott, the divers' rigging foreman (whose sooty, blood-flecked face he had passed earlier) came up to him.

'What happened?'

'I was on the landing stage away by the gritting-shack. Then a bleedin' great fireball flashed over us heads. Before I knew it, I was smack flat on the ground.'

Elliott insisted he was OK and pointed out the button which Punchard was looking for. The compressor whined a little more and then died out.

It was crucial for Punchard and Carroll to get up the next two flights of stairs to their lifeboat station, and despite being reminded that their path was blocked, they headed up once more. At the top of the stairs they found that the 85ft-level now had a layer of thick black smoke, making their eyes sting and water. They moved carefully along the walkway of the west face, holding tightly to the handrail, and made for the stairway that led up to the control room. Although they could feel the heat, they could not yet see its source. Only when they had reached the middle of the walkway and looked up did they realize that the smoke was masking a ferocious fire with fingers wrapping around the pedestal that supported the rig's crane. Up on the 107ft-level, the men's lifeboat was visible through the smoke and flame but impossible to reach.

They turned round and headed back to the north-west corner, where the small pocket of platform which was clear of smoke had shrunk. This was unfortunate, as the number of men occupying it had increased. Earlier in the day an area had been taped off for repairs to the grating. Now Punchard ducked under the tape, jumped up and down to check the grating's stability and then removed the tape in order to provide extra space. Derek Ellington, a fitter with the Wood Group (one of the many contractors on the platform), was talking about the noise he had heard prior to the explosion. He said it was a loud, high-pitched scream 'like a man being gripped by his balls'. Among the group there was frenzied discussion on what to do. Voices were raised and curses rang out as the seriousness of the situation was brought home by the crackle of flame and the acrid stench of smoke. One man said they should wait for the control room to issue instructions, while another maintained that this was not going to happen.

'You should see the fuckin' state of the place. There's wire hanging down everywhere. Even the bloody bulkheads have buckled in.'

When asked how the man knew, he said he had just come from there: 'The explosion blew me straight across the fucking room on my chair.'

Smoke began to wrap around the group and Punchard grew increasingly anxious, checking the guard rail to see how quickly he could squeeze through the three bars in the event of an escalation, which seemed increasingly likely as fine, fiery embers were now mixing with the smoke. Then a loud, high-pitched scream began. Then stopped. Then began again. Punchard thought it must be escaping gas, and the group decided to head down the stairs back to the 68ft-level.

They found that the smoke there had become much greater, as had the number of men gathered at the corner. Although the divers were now all reunited, the news was not good. Parry-Davies said the 'Wendy Hut' was now entirely ablaze, as were the nets holding up the umbilical; one net had already broken away and fallen into the sea. Christopher Niven found a storage box containing life jackets and began to pass them up the line, but there weren't enough to go round – thirty for a crowd numbering over forty. Once this was realized, a few divers who had received a jacket took it off and then passed it to someone else. They at least had experience of swimming in the North Sea, while some of the other men couldn't even swim. The life jackets were orange and non-inflatable; they passed over the head and had stiff flotation compartments at the front and back, reflective stripes and a whistle.

It was clear to Stan MacLeod and Barry Barber that if they

couldn't climb up to the lifeboats they would have to scramble down. A few yards from the corner, attached to the railings but hanging over the sea, was an emergency life raft sealed inside a white plastic capsule like a small, fat man's coffin. Keith Cunningham set about launching it. First he removed the steel pin that secured the safety straps and watched as the raft, still sealed inside the plastic capsule, tumbled down into the sea. A line which tethered the capsule to the railings was then to be pulled in order to activate a gas canister inside the raft, causing it to inflate and burst out of the capsule.

Unfortunately, although the capsule had already fallen 68 feet, Cunningham was unaware that a further 68 feet of rope still had to be pulled out until the line went taut and the canister activated. No matter how long and hard he tugged at the raft it refused to inflate. At one point it seemed that the line was taut and the capsule was now being hauled out of the water, though it might have been carried up on the back of the waves (a 5ft-swell having recently blown up). Frustrated, he abandoned his attempts when the waves carried the capsule under the rig and so rendered it inaccessible even if it had opened.

Cunningham then turned his mind to providing a communal means of escape. Beside the emergency life raft was a knotted escape rope, which he began to unwrap and lower down towards the water. This was designed as a last resort, for it meant going into the water (which even in summer remained just ten degrees) unprotected by lifeboat or raft. The act of lowering the rope sent a frisson of fear through the crowd of men who were bunched tightly together in boiler suits and orange life jackets. Escape now involved going hand-over-hand down a thick, coarse woollen rope, suspended almost 70 feet above a sea whose temperature, they had been repeatedly told, would kill them within minutes. A few men began to back away.

The first man to go down was Edward Punchard. He was thin and wiry, with dark hair and an attitude that could best be described as aggressively positive. As a result of his diver training he was also fitter than he had ever been in his life and so volunteered to source a suitable escape route. Although he tested the rope before climbing over the barrier, as soon as he swung out he and the rope suddenly dropped 10 feet before catching. Swinging violently from side to side with his feet kicking in mid-air, he nevertheless managed to hang on, and as the rope's pendulum motion settled to a stop he caught his breath and then scrambled down through the smoke to the spider deck, a narrow metal mesh walkway that skirted the platform just 20 feet above sea level and allowed access to each of the platform's ten steel legs. Looking up at the underside of the platform, Punchard saw the bright orange flames dance at one corner and send billows of dark, soot-grey smoke pirouetting into the night sky. He wasn't going near the legs closest to the flames and so he followed the gangway down to leg B5, but as far as he could see it didn't have a vertical ladder. However, further on stood B4, which did have a short ladder leading down to a boat bumper, a circular steel tube like an upside-down letter 'L' that projected from the leg and went down into the water as a barrier to protect the leg against careless captains and errant vessels. The fire had illuminated the sea, turning it from a forbidding grey into a warm gold. A few hundred yards distant was the thin grey Fast Rescue Craft (FRC) from the *Silver Pit*, the converted fishing vessel whose job was to stand sentinel over the rig in case of emergencies. Shaped like an arrowhead and crewed by four men in bright orange survival suits, the FRC could manoeuvre at high speed and was designed for close-quarters rescue.

Punchard waved to the boat, pointed to the B4 leg and

watched as the FRC changed course and headed his way. It was, he thought, as easy as hailing a taxi.

Back at the 68ft-level, Stan MacLeod and Barry Barber had organized the men into a line next to the escape rope. As soon as one person climbed over the barrier and began to scramble down, MacLeod would tap the next one on the shoulder and say: 'Right. Go!' As the men queued they remained largely silent, standing alone with their thoughts and struggling to suppress the mounting anxiety which intensified with the noise from the fire. Standing in line, Alastair Mackay, a diving systems electrician, thought the sound resembled 'the whole platform . . . dying'.

For the first few minutes the line moved smoothly, although those in life jackets found the climb difficult as the thick knots repeatedly stuck under the stiff flotation compartments on the front of the orange vests. A few couldn't move when tapped on the shoulder, their legs fused by fear. Instead they stepped aside and allowed the next in line to go. However, one responded to MacLeod's bullying and did manage to climb over. Unfortunately another worker, once on the rope and a few feet down, became petrified, suspended 60 feet above the grey waters. Unable to move down and unwilling to move back up, he held up the line until MacLeod climbed over, lowered himself down and sat on the man's head while screaming at him to shift. He shifted. Yet once on the rope MacLeod had no choice but to continue on down himself, leaving Barber in charge of the remainder of the impromptu evacuation.

The majority of men scrambled down the rope to the rig leg where they climbed on to the FRC in groups of four or five before speeding back to the *Silver Pit*. Those pushed up against the metal railings on the 68ft-level were soon driven down by encroaching heat, fire and smoke to an even smaller navigation platform that hung suspended below the north-west corner. It was from here that Dick Common, a site administrator, swung out on to the rope, which was now greasy with the sweat of dozens of clammy hands. After lowering himself for a short distance, he felt his feet begin to lose their purchase and he adjusted them, but in the process felt his hands slip. He barely had time to scream as he fell from 40 feet, struck the boat bumper and splashed into the water. Shocked, his colleagues above began to yell, shout and wave their arms in a desperate attempt to attract the attention of the FRC crew in the waters below. They then pointed repeatedly in the direction where Common had splashed down, and were relieved when the craft changed course and headed towards the spot.

Once Keith Cunningham had climbed down to the 20ft-walkway he stayed there to hold the rope and assist the older guys, men in their late forties and early fifties. Some were a little overweight, and frightened to be swinging out over the sea with nothing but a rope and their own strength to support them. He held on until Barry Barber came down, thinking he was one of the last. Just two men now remained, seemingly too frightened to move as they crouched down together against the railings of the navigation deck. Cunningham kept looking up at them and shouting for them to come down, but they didn't move and eventually were covered in a shroud of smoke.

Then Cunningham heard a tremendous din, a noise so loud it seemed to shake him to the core. He spun round and was able to utter only a single word to Barry Barber, who was also standing on the walkway, before immediately launching himself into space.

'Jump.'

A second massive explosion had occurred which, like the instant inflation of the canopy of a large hot-air balloon, sent a ball of curling white and orange flame along the underside of the rig and into Cunningham's path.

He appeared to fall through flame and then sink ten feet or so into the sea's chill. Looking up from under the surface as if through a murky pane of glass, he could see a bright light and feared surfacing, but had no choice but to rise and breathe. Once he broke the surface he saw, and felt, that the sea had been sprayed with oil and bright yellow flames surrounded him. The air above the water had become superheated to over 200 degrees and was, he thought, like putting your head in a roasting oven. He took a deep breath, dived back down, and once under the water began to kick away from the rig.

John Barr had been climbing down the rungs of the ladder towards the sea when the explosion sent men racing down after him, pushing and shoving to clear the way to the icy water and a respite from the heat that was prickling and burning their skin. In the water a total of six men, including Barr and Christopher Niven, were struggling to put the steel boat bumper between them and the intense heat, clinging to the steel pipe and pressing their faces against it as a protective barrier, then ducking their heads under the water every few seconds to keep cool as their faces began to burn. They looked up as chunks of burning debris, some the size of a small car, fell into the water. The FRC retreated in the face of the flames. On-board was Richard Common, who had been dragged shaking and spluttering over the side of the craft a minute or so before the blast and was now wedged on the floor, peering out through tiny portholes at the flames and convinced that at any second they would all be engulfed by the fireball.

At the boat bumper, the men found that the waves had such force that simply holding on to the ladder required all their strength. Chris Niven was carried up and down by the swell. He looked out and saw Barry Barber about fifteen feet to the west, struggling to hold on to a steel strop. It was clear that Barber had no protection from the heat and Niven thought: 'Hell, what's he doing there? Can't he swim towards us?'

Then Niven was again carried up and down by the waves. The next time he looked Barber was gone.

•

On the deck of the *Silver Pit* the first few rescued men stood and watched, unable to quite believe what they saw. The fishing vessel was stern on, 200 metres from the south side. The wind blowing the thick black smoke away to the north revealed that flames ran almost from the surface of the sea, up in a thick band under the platform, enveloping it, then roared on up around the drilling derrick before flourishing three or four hundred feet above the rig's highest point and into the night sky. A solid wall of heat, almost tangible, now butted against their faces. The rope handrails were smouldering. Through the heat-haze shimmer, they could glimpse figures on different levels of the platform, a fraction of the 200 men still on board.

It was 10.20 p.m. Twenty minutes earlier Gareth Parry-Davies had been under the surface. Now he stood on the ship's deck, still in his diving suit, and looked out at Piper Alpha. 'I can recall this terrible feeling – I had this feeling that the platform was finished then, that nothing was going to put that fire out.'

PART ONE

ONE YEAR EARLIER

1. PORTRAIT OF AN ARTIST
AS A YOUNG RIGGER

'Small' said the man behind the counter as he looked the young woman up and down before heaving the orange rubber suit onto the counter top. She smiled, struggled to pick it up and then moved over to the benches to change. It was just after 6 o'clock in the morning and the helicopter hub at Dyce Airport in Aberdeen was packed with unshaven, scruffy men carrying duffel bags and the scent of last night's spirits. As alcohol was forbidden offshore, with each man's bag searched for secret supplies, it was traditional to begin two weeks' abstinence with one final night of abundance. She thought that some men were beating the ban by carrying alcohol on board still coursing through their veins. They weren't drunk enough to be ejected by security, but she certainly wouldn't trust them to drive. Outside, the men's various 'rides' queued to lift off from the world's busiest heliport. Pumas and Sikorsky S-61s, draped in the red, white and blue of the Bristow company, sat alongside the large white Chinooks of British Airways. Today she was flying in the smaller chopper of British Caledonian helicopters with its livery of gold, white and blue. If, that is, she could ever get into the survival suit.

In the early days of the North Sea oil industry, when companies sped offshore leaving safety regulations struggling to catch up, men would wear jeans and jumpers, or even a suit on the flights out to distant rigs and platforms. These were journeys where there is no place to land in case of emergency except a sea whose temperature rarely rose above ten degrees, even at the height of summer. Today, the stiff, rubber-lined survival suit with its bewildering array of zips and flaps was the mandatory dress code. Like a baby's romper suit, it was an all-in-one with bootees over which you put your own shoes. First you had to step into the lower half, then squeeze your head through a neck-hole as tight as a strangler's grip and force your hands through wrist seals like handcuffs. On the right shoulder, like a small one-eyed parrot, sat a strobe light which, in the event of ditching, was to be fitted onto the accompanying hood.

Once she had completed the sartorial obstacle course designed to ensure her survival, she waited for the flight to be called. She was one of only two women in a room full of tired, bleary-eyed men whom she then followed out onto the tarmac, up the steps and into the cabin, ducking instinctively although the rotor-blades were at least six feet above her head. There was a central aisle with one line of single seats on either side, and she took a seat towards the back before putting on the orange earphones provided. The din of the engine and rotors, already loud and making conversation impossible, rose to the point of pain in preparation for take-off. The helicopter then began to rock before rising straight up a dozen feet, the nose dipping down slightly as the craft swung away over the grey concrete of the airport and the granite terraces of Aberdeen. In a minute the grey of the city was replaced by the green and brown of passing fields, the brief white strip of beach whizzing past before the view was embraced by the blue of the North Sea.

Familiarity may not breed contempt but it nurtures indiffer-ence, so the men on board ignored the view and, instead, tilted their heads back to sleep or forward into the folds of a newspaper or pages of a book. But she preferred to look out of the window and down on to the water speckled with white cresting waves. When the helicopter rose into the clouds, she pulled out a Pentax MX camera and focused the lens on the bearded, balding man across the aisle. The shutter closed with a click, soft enough not to wake him from his slumber. The men on-board the early-morning flight to Piper Alpha were on their way to work, and so was she.

With her hair dyed orange and spiked, the only similarity between the artist Sue Jane Taylor and the 'bears' who worked offshore was an appreciation for warm woollen knitwear. A fine art graduate of Gray's School of Art in Aberdeen, Taylor went on to complete a scholarship at the prestigious Konsthogskolan art school in Stockholm, after completing postgraduate work at the Slade School in London where, in the summer of 1987, she was currently employed. Born in Dingwall in 1960, she was raised with her elder brother and sisters amid the remoteness of the Black Isle, in the small village of Munlochy, where her father owned a Spar franchise and kept bees.

As a child, Sue Jane was as interested in local history as she was in her fledgling experiments with oil and watercolours. Like the rest of her community, she watched in amusement and awe as the landscape was transformed when the Nigg construction yard on the Cromarty Firth opened in the early 1970s and began rolling out the 'cathedrals of steel' as the rigs and platforms were soon christened. She was not the first family member to pursue art. Her elder sister, Romy, had attended Edinburgh College of Art before withdrawing after a breakdown which was later diagnosed as schizophrenia. It was an experience that prompted Sue Jane to stay closer to home for her studies, favouring Aberdeen over Edinburgh. There, in pubs like the Prince of Wales, she and her fellow students learned to recognize the offshore worker by the cut of his designer jeans, the scent of his strong cologne and the addition to the usual gents' wardrobe of cowboy boots – usually two sizes too big – and an ostentatious cigar.

Art and the offshore world drifted together for Sue in 1984 when Peter Wordie, chairman of the Stirling Shipping Company and a passionate art collector whose father had been a geologist on Ernest Shackleton's expedition to the South Pole between 1914 and 1917, invited her to take part (with other artists) in celebrating the company's tenth anniversary in the lucrative work that lay behind the horizon.

Five days in the depth of winter on a cramped supply vessel running out to the rigs might have left other artists pitching in their shared cabin and dreaming of pastoral landscapes bathed in a warm and golden light; but for Sue, her time spent on-board the *Stirling Teal* was to prove an epiphany. The experience of drawing up at midnight to an oil platform's massive steel legs and watching as, under spotlights, the crane driver timed the descent of his hook with the *Teal*'s rise on the wave crest, served to open her eyes to a world where nature and industry collided and squeezed men to their limits. Surely this was as deserving of artistic representation as anything on land, even more so given the public's ignorance of what is required to enable them to fill up a car or boil an egg. Three large etchings emerged from the trip, but as Sue sailed back into Aberdeen harbour it was with the ambition to take the industry's impression in oil, sculpture and paint.

While working as a teaching assistant at the Slade School, Sue would contact the headquarters of oil companies and request permission to visit their sites, using the £5,000 presented by the Glenfiddich Living Scotland Award (a grant to support artists in their work) to cover her costs. In August 1986 Occidental arranged for her to visit the company's oil terminal on the

Orcadian island of Flotta, and now, a year on, she was en route to spend five days on the company's most profitable platform.

The helicopter dropped down below the clouds and Sue could look out on the sea again and spot the great haddock trawlers now reduced to toy boats in a bath. Then, as her eyes scanned along the surface, she could see in the distance a structure rising from the sea, like a distant city on the edge of a vast plain. She couldn't help but feel a frisson of excitement. It was, she thought, like coming in to land at a spaceport. She could see men walking around the steel structure as the helicopter circled and dropped down to land on the helideck, which from a distance resembled a 50 pence piece on an outstretched palm.

Sue shuddered as the helicopter touched down, then unfastened her seat belt, packed away her camera and followed the men out of the door. The wind blew in her face. It was a bright, clear morning, not yet 10 o'clock, and she stopped to take in the view from the most expensive real estate on the planet. In the air seagulls circled with cormorants, while in the waters below the standby vessel bobbed on its perpetual voyage, like the minute hand on a clock, round and round the platform.

The platform itself was a bewildering maze of metal parts which at one angle appeared futuristic, while at another seemed strangely medieval in its raw, unvarnished functionality. The derrick resembled a lighthouse peeled of its stone skin to reveal the steel skeleton beneath, while the twin flares reached out from either side of the platform like a greedy child's outstretched arms.

A dark-haired slightly balding man, with relaxed shoulders and a pleasant face framed by glasses, came up to meet her. Colin Seaton, a 51-year-old Yorkshireman, was the platform's Off-shore Installation Manager (OIM), the 'captain' of the platform, with authority over everyone on-board and all oil and gas operations. A veteran employee with Occidental, he helped Taylor out of her survival suit and led her downstairs to 'A' level, where he had his office and a cabin had been reserved for the duration of her stay. The cabin was small, but not uncomfortable, with two bunk beds pushed against one wall, a small chest of drawers and a shower and toilet area. The small porthole window offered a sea view – as did every window on the platform – and came equipped with a colourful set of orange curtains.

After unpacking her inks, sketchpads, Conté sticks, camera and tripod, Taylor headed back upstairs where Seaton had arranged for her to lunch with the platform's medic, Gareth Watkin, and Trisha, the only other woman on the flight, who was in charge of statistics for safety standards.

The canteen – or galley as it was known – was on the top floor, directly beneath the helideck. It was a long rectangular room; along one wall ran windows draped with gingham curtains, through which light flooded to the benefit of the potted plants. Two large chill cabinets contained an array of cakes, trifles and individual fruit salads. Along the opposite wall, the hot food service offered a banquet of steak, roasts, fresh fish, an assortment of potatoes and heaped bowls of salad, all freshly prepared in the kitchen which lay behind. Taylor collected a tray, chose the fish, then joined the pair at one of the two dozen formica-topped tables, pulling up a metal chair with a cushioned seat.

Watkin was tall and thin, with a greying moustache. A former merchant seaman, he had 'found God' and used his faith to see him through each two-week shift and the grisly accidents which he occasionally had to attend. Looking around the room, Taylor considered it quite a pleasant place to eat. She glanced over at the men queuing for their food, then looked up and smiled. Above the kitchen gantry was a piece of glass on which was engraved a highland piper in full skirl. Nice touch, she thought.

After lunch she was passed on to Jack Patience, the head safety officer, who led her down to the lifeboat to which she had been designated during her stay. Safe in the knowledge of how she should get off, it was time for the grand tour so that she would understand, or at least begin to comprehend, exactly what she was on.

Colin decided to start at the equivalent of the 'coal face' where the dirty work was done. From his office the pair set out across the narrow gangway that covered the pipe deck and then headed down to the drilling floor where they stood beneath the drilling derrick, the tip of which, he explained, stood 289 feet above the sea. It was here, he informed Sue, shouting over the din, that the muddy, physical act of oil extraction began. It was here, as she could see by the men moving around her, that the 'roughnecks' – the assistants to the driller – toiled with their overalls, gloves and boots, quickly becoming coated in the pungent, greasy drilling 'mud'. The 'mud', which Colin scooped up in a gloved hand to show her, was in fact an artificial construct of different chemicals. 'We use it to cool the drilling bit and prevent the hole from collapsing before it can be lined with steel pipes,' he shouted over the noise.

During the past decade 36 separate wells had been dug from the drill floor down into different reservoirs. While on a diagram each hole radiates neatly away from the rig, the reality is that 9,000 feet below the seabed lay a jumble of pipelines which Colin said had the clean precise lines of a plate of spaghetti. Using computer-controlled directional drilling, the 'toolpushers' – drillers – are, or so it was joked, actually trying to autograph their names underneath the platforms while drilling each well.

When drilling a new well – or deepening an old one – the roughnecks regularly have to perform a 'round-trip', which involves pulling the entire 9,000 feet of drill string back up the hole to change the drill bit, described by one poet as 'like a great clenched fist with cogged teeth where the bent fingers should be'. The pipe is hauled up in lengths of 30 feet, three lengths at a time, using a large hook and swivel attached to a wire rope that runs up to the top of the derrick, over a pulley and back down to a giant drum on the drill floor. For the twelve hours this journey can take, the roughnecks clamp tongs round each three-length section of five-inch pipe while a rotary table spins and unscrews it. The pipe is then leaned against the side of the derrick.

Colin explained that the temperature in the oil reservoir was around 200 degrees Fahrenheit, and that the pipes emerge still steaming with heat. Once the 'round trip' is complete, it is tradition for roughnecks to spit down the hole. When Piper Alpha embarked on drilling operations the entire platform would vibrate, sometimes violently enough to topple a tea cup.

Bidding goodbye to the crew, and promising to return to document their work, Sue and Colin then left the drill floor and headed for the module deck, a series of closed areas beneath the drill deck where the majority of the platform's machinery is stored. The next stop on the tour was what happened after each new well had been drilled. He led her into A Module, a low open room where Sue looked at a series of 36 pipes which rose out of the steel floor in three ranks of twelve, each one topped with a complicated contraption resembling a large tilted steel cube. It was painted silver with red and yellow pipes sprouting from the side, out of which grew various gauges and dials. Colin explained that these were known as the 'Christmas trees'.

Although the level of detail was becoming overwhelming, her guide had such evident pride in his platform that Sue persevered, nodding along when she should, if she was honest, be shaking her head. Standing in the midst of the room, Colin looked out at the forest of 'trees' and explained that after a well has been drilled, and the reservoir tapped, the hole is then lined with casings cemented into the rock. The pipe – or conductor as it is known – that runs from the well head on the platform down through 474 feet of seawater, then a further 300 feet of seabed, is one inch thick and 30 inches in diameter; but the deeper the well the narrower the pipe casing becomes, until the piece that punches into the reservoir and draws up the oil at the beginning of its long journey to the surface has a diameter of just seven inches.

She listened as Colin went on to explain that the coloured pipes on one side of the Christmas tree control the flow of the oil and gas from each well, while those on the other side are 'kill pipes' put in place to shut down the well should any problem occur. This was achieved by pumping into the stricken well a high-density drilling mud that was heavy enough to push down the pressure in the well which brought the oil and gas to the surface.

The pair then followed the pipes north and into B Module to the process module, a refinery of silver pipes and large cylinders in which the gas was separated from the oil in a process that left Taylor's head spinning. At the west end of the module she was shown the Main Oil Line (MOL) pumps. The next room, C Module, was the gas compression module, where at the east end sat three centrifugal compressors, each in its own separate enclosure and equipped with a turbine.

At the north end was D Module, where the power for the platform was generated by two giant John Brown generators whose massive exhausts projected out of the north side to the platform like two steel garbage chutes for giants. Up above, on a mezzanine level, sat the control room through whose glass-fronted doors she had yet to pass.

•

The tour had been conducted while Sue was wearing trainers, and she had repeatedly slipped on the metal grille walkways which were still slick from the recent rain. When she returned to Seaton's office, Jack Patience called the company's office in Aberdeen and asked for a proper pair of leather, steel-toe-capped boots, in the correct size, to be brought out on the next flight. Meanwhile Colin Seaton, having listened to what Taylor hoped to achieve, agreed that rather than be tied down by a chaperon she could go where she liked, whenever she liked, a concession few other OIMs would even consider. But not until she was suitably shod, he emphasized.

While waiting for the flying boots to land, Taylor wandered round the interior, slowly familiarizing herself with the various accommodation blocks which over time had been bolted together and connected by a labyrinth of corridors. To the new arrival each corridor appeared identical, with the same dull grey carpets, blue-washed walls and repetitious rows of cabin doors, only the small numbers providing any sign of individuality. She stopped and took a picture at the bond shop, a tiny kiosk barely big enough for its moustached attendant, but where the shelves were stacked with chocolate bars, cigarettes, cigars, perfume and packets of Fisherman's Friends.

At 28, Sue was an attractive woman and received enough furtive glances and broad smiles from the men to ask a group of workers, waiting in the despatch area for a delayed flight, what they thought of women working offshore. No one was in favour. While most insisted that the work (heavy, dirty and exhausting) was unsuitable for a woman, a few also explained that their wives wouldn't like the competition. The 'competition' was reduced by one when Occidental's safety officer departed around teatime, leaving Taylor as the only woman trapped on a steel island with 160 men. Tony the chef, a gregarious character with a wrinkled face partially hidden by thick silver spectacles, took delight in pointing this out when she returned to the galley for the evening meal. In his white hat, pristine chef's jacket and striped trousers – an ensemble set off with a jauntily angled blue neckerchief – Tony was among the smartest-dressed men on board. He was also among the most forward, quickly dubbing her 'sunshine' and regaling her with the tonnage of beef, fish, potatoes, vegetables and other kitchen supplies that required to be shipped out each week in order to keep the men well fed. He said it used to be better in the early days, when T-bone steaks the size of a plate were served up nightly, but oil prices had slumped and there was cost cutting across the board.

Offshore life, as Taylor learned, hinged on a set routine. At the end of a twelve-hour day shift, which usually finished at 6 or 7 p.m., men would shower and then head to the canteen for dinner. Afterwards the recreational options were limited. The games room was equipped with table tennis and a full-size snooker table on which breaks of one hundred plus were not uncommon, given the practice some players put in each night. There was also a small library that operated a 'one in, one out' system, requiring workers to return a book before they could take one out, with Westerns roaming all across the shelves, followed by detectives. Porn mags were openly exchanged and deemed more acceptable reading material than the union literature, which was treated furtively by comparison.

The cinema, which could accommodate sixty in tiered seating, showed a different movie each night, including hard-core sex films. These were illegal in Britain, but brought in by Norwegian workers and popular with the men, who reserved their places by leaving a banana or a book on their seat of choice. Taylor was relieved to discover that the evening's film was not *Debbie Does Dallas* but *84 Charing Cross Road*, the story of an epistolary friendship between two middle-aged bibliophiles which, unsurprisingly, had failed to prompt a rush or even a single 'reservation' banana. When she looked in, the room was half empty.

As Gareth had arrived bearing her new boots, presenting them as she dined, Taylor decided to walk outside for some air before retiring to her cabin at around 10.30 p.m. As she lay in the dark she could feel the bunk bed vibrate slightly, driven by the turbines and generators below, whose muffled noise still managed to squeeze through the soundproofed walls of the cabin. She fell asleep, breathing in the strange scent of oil and chemicals that perfumed the air, and pondering pipelines to the centre of the earth.

Sue Jane Taylor was determined to amass a bulk of material for use in different artistic mediums, so the next few days were packed with activity that involved clambering the breadth, depth and height of the platform. Memories of the brief voyage on the *Stirling Teal* came back as she reversed her role from that winter's night and looked down from the safety of the platform on to the supply vessel, over whose deck the waves repeatedly washed as the crane lowered down its hook and hauled up crates, some the size of portacabins. Gareth – who, when free, acted as her guide – then took her underneath the platform where a network of staircases and gangways led down towards the sea. Since the sea was riding high and washing over the 20ft-deck, the pair restricted themselves to the 40ft-deck, which still gave Taylor an impressive view of the well conductors, a row of broad steel pipes that rose up from the seabed, carrying oil and plugged into the floor of the modules above their heads.

She joined Derek Hill in the crane cabin and he allowed her to take the controls and load on to the supply ship a small package of duty-free cigarettes ordered by the ship's crew. She listened as John Wakefield, an instrument technician from Grimsby, explained how troops from the Special Boat Service regularly rehearsed taking control of the platform in the event of an attack by the IRA or foreign terrorists. He said they could clamber up the side of the structure like black beetles in their camouflage gear, and secure the control room inside 40 seconds.

It took another day for Sue to gain access to what the SBS achieved in less than a minute. The control room of Piper Alpha reminded her of the flight deck of a spaceship with its bank of dials and lights, digital readout and rows of computers, some of which appeared quite dated, and television screens that displayed an unreadable language of numbers, diagrams and abbreviations. Colin Seaton explained how the engineers could monitor each function in the production process and, when drilling, know the exact position of the drill bit 2½ miles under the earth. The contrast in this clean, sterile room to the dirt and mud of the drill floor seemed stark. Again she pondered the idea of two separate worlds, one science fiction and the other medieval, inhabiting the same square one-acre platform.

One afternoon she positioned herself in just the right spot to enjoy the heat off the flare boom while she sketched details of the platform in Conté and inks. Looking down, she saw a seal teasing a seagull just a few yards from one of the platform legs. The seal would leave a fish floating on the surface, but just as the gull dive-bombed for lunch would snatch it away, repeating the process for its own amusement. Then she noticed the thin slithery line of an oil slick a few yards from the seal. Oil companies, she was later told, were permitted to discharge five barrels of crude oil per day, a limit they regularly exceeded. When she expressed her surprise to one worker, he explained that news of the many accidents and mishaps offshore never reached the mainland. Piper Alpha, he said, was an old rig.

Two nights later, Sue retired to bed early but rose after midnight and climbed the stairs up to the helideck. The weather was dry but windy, and the deck was deserted, with the only sound the light thrum-thrum of machinery down below. She looked out and saw that sky and sea had merged into a single dark screen on which she spotted the neon lights of Piper Alpha's sister platforms, Claymore and the Tartan, while closer still were the flickering lights of the safety vessel in floating vigil below. Sue had been experimenting with infrared film and angled the camera over the side, where the waves lapped against the well conductors. During periods of her stay she had begun to see the whole process of extracting oil and gas from the depths of the earth with the same glassy-eyed indifference as the veteran rigger, but in the darkness the isolation of the men was tangible. If, she thought, she was to fall over this instant, no one would see or hear her, and even if the standby vessel spotted the splash, would they make it in time? What then of the platform's population?

A rural upbringing had taught Taylor respect for the land, and in the darkness she reflected on the powerful forces the men here had dug out, corralled and imprisoned in pipes: an achievement worthy of Prometheus.

Then she crouched down low and photographed the platform's name stencilled in tall white letters, 'Piper – A', which from her angle seemed to stretch out into the night.

When Sue lifted off from those letters at the end of her stay, she did so with more weight in her luggage that when she had touched down. Piper Alpha had passed 100 days without an accident, and staff had been presented with a pewter tankard on which was written: 'We did 100 days. Piper A'. This was, she thought, a souvenir worth holding on to.

2. 'GOD'S LAST CHANCE
FOR THE BRITISH'

The path to Piper Alpha ran through Occidental, the multinational oil company controlled by Armand Hammer: a man of contradictions. A capitalist friend of communism, he claimed to be the only man to have known both Vladimir Lenin and Ronald Reagan. A patron of the arts and collector of Impressionist and post-Impressionist paintings, he was described as 'a man who vulgarized everything he touched'. A doctor who graduated from Columbia University, he never practised but instead used his intelligence to operate on the stock market and deployed his considerable bedside manner on dictators, presidents and princes. Born in Manhattan, he claimed to have been named after Armand Duval, a character in *La Dame aux Camélias* by Alexandre Dumas, although later biographers insist that his father, Julius Hammer, had the 'Arm and Hammer' symbol of the Socialist Labour Party firmly in mind. A committed socialist, Dr Julius had arrived in New York from Odessa in 1875 and built up a medical practice and a chain of five pharmacies, a business which passed into the care of his sons when in 1919 he was sentenced to 2½ years in Sing-Sing after performing an abortion on a woman who later died.

In 1921, Armand Hammer was sent to the Soviet Union by his brothers to retrieve $150,000 owed on a consignment of drugs shipped a few years previously. He had been about to begin an internship at Bellevue Hospital, but the wards gave way to board-rooms as he struck a deal with Lenin to exchange surplus supplies of American wheat for Russian caviar and furs. He had smoothed the path into the new Soviet Union by bringing medical supplies to assist against an outbreak of typhus, and then based himself there while building up an array of businesses including the manufacture and exportation of cheap pencils. The contacts and friendships built up over chilled vodka, pickled herring and solid profits were to benefit him in later life as he straddled two oppos-ing worlds, one foot in both camps of the Cold War.

Dr Armand Hammer, as he would insist on being named, was elected president and chief operating officer of Occidental Petroleum in 1957. Four years later, under his stewardship, the company discovered the second-largest natural gas field in Cali-fornia, at Lathrop in the Sacramento basin, a bonanza that helped to fund the firm's expansion into Libya in 1965, where under the benign watch of King Idris they enjoyed favourable terms in the exploitation of the country's rich oilfields. The bloodless coup in September 1969, which placed Muammar el-Qaddafi – a 27-year-old colonel in the Libyan army – in control of the nation, raised the fear of nationalization or, at the very least a major hike in the price per barrel which Occidental paid the country. While Dr Hammer immediately cancelled the regular payments to the Swiss bank accounts of King Idris's acolytes in the oil department that had long greased relations, and instead began to source the appropriate men in Qaddafi's camp, he turned his attention to discovering alternative sources of oil where conditions were politically more stable.

•

Oil takes time. The process that laid in place deposits of hydro-carbons (the term for oil, tar and natural gas) deep under the rocks of the North Sea developed over hundreds of millions of years, for locked inside Scotland's geography are among the oldest rocks on the planet, stretching back 2,500 million years. What happened was a fusion of heat, stone and time which laid layer upon layer of various rock types, building up from the ancient past to the (in geological terms) almost present.

At the bottom level were crystals and sand, the product of volcanoes, which were melted and fused and then smoothed into granitic rocks ten kilometres thick. This was followed by further layers of sandstone, then blankets of schists and quartz. Scotland then lay on the south-eastern rim of one continent which was then forced by the rotating of the earth's tectonic plates into what is now England and Wales, when this land mass lay in the southern hemisphere.

A shallow sea then spread over the area 370 million years ago, populated by plants and creatures whose death and subsequent decay formed into seams of limestone. These carboniferous jungles were later compressed under more rock to form coal seams. Dinosaurs arrived 200 million years ago, and for 150 million years passed through what were known as the Variscan mountains and a landscape of shallow seas and lakes. It was the vegetation and plankton of these seas and lakes, decayed and compressed beneath Jurassic sandstones, which transformed into oil and gas.

Volcanic activity moved up the western ring of Scotland while water flowed into what would become the North Sea. A final eruption of volcanic activity 58 million years ago birthed new mountains, only for them to be worn down by glaciation and further erosion, whilst Scotland, with the rest of Britain, slowly spun from the southern hemisphere to the north like a dancer in a waltz.

What would come to be described as 'God's last chance for Britain', the billions of fragments of decayed organic matter, was then entombed by thousands of feet of sandstone strata, mud and eventually hundreds of feet of salt water. During the long eons in which hydrocarbons form, the pressure from the rocks and strata above squeezes them out of the mud in which they once lay and forces them into the gaps and holes of more porous rocks such as sandstone, chalk or limestone. Unfortunately for oilmen, it is a myth that they pool in massive underground caves, waiting for liberation when the drill bit hits. The reality is more complex. They squeeze between the grains of 'reservoir' rocks, gather in a network of water-filled pore spaces and permeate upwards. They then become locked in place by a 'cap' rock through which they cannot penetrate.

Reservoir rocks and cap rocks are known as 'traps' and only occur in sedimentary rocks of the Cretaceous and Jurassic periods, usually at a depth of between six and 12,000 feet, six times the depth of the deepest mine.

Oil and man do mix, though not without difficulty and discomfort and too frequently death. The inflammatory property of crude oil has been known for centuries, courtesy of oil seepages from rock. Peat soaked in oil was used as firelighters, while miners scooped the viscous fluid from rock walls and used it to lubricate their cartwheels. Hugh Miller, the pioneering geologist, wrote in 1854 about the property of oil claimed from the Moray Firth: 'Every heavier storm from the sea tells of its [oil's] existence by tossing ashore fragments of dark bituminous shale. The shale is so largely charged with inflammable matter as to burn with a strong flame as if steeped in tar or oil.'

Before the wisdom of geologists righted the wrongs, a number of theories rose about the origins of oil. One philosopher believed the earth was an animal, water its blood, rock its bones, and that like the whale, when man bored through the earth's skin oil emerged like blubber. Another and even more obscure belief was that it was the urine of whales carried from the North

Pole via subterranean channels. Yet the early 'wildcatters', the American pioneers of oil exploration, when confronted by fountaining geysers of black oil said it was as if 'the earth had cut an artery'.

•

Britain's oil exploration expanded into the Far and Middle East with the D'Arcy Exploration Company discovery of an oilfield ten miles wide in Persia. In time, the company would become Anglo-Persian Oil and fall under the control of the British government when Winston Churchill, then First Lord of the Admiralty, wished to secure supplies for the fleet that he had switched from coal- to oil-fuelled engines. Britain's first commercial oilfield was tapped in 1939 at Eakring, a village in Sherwood Forest. A gang of American wildcatters from Oklahoma sailed over on the *Queen Mary*, with their drilling equipment carried on naval vessels, to teach their future allies and assist them in accessing 100,000 barrels within a year, a target they easily exceeded by going on to produce 110,000 tons of oil, which was of crucial importance during the war years.

Between 1949 and 1972 demand for oil would increase 15-fold and result in the rise of the 'Seven Sisters', the world's most powerful oil companies – five American and two British.

The key that opened the door to exploration and extraction from the North Sea was the discovery in 1958 of a giant gas field, the largest outside Russia, on the north coast of the Dutch province of Groningen. The oil companies reasoned that if there was gas there, might there be gas and perhaps even oil further out in the North Sea? Geologists had long surmised that hydrocarbons might lie in the deeper waters further north, where the rock formations of Yorkshire – which do bear gas – dip under the sea, only to spring up again in the gas fields of north-west Europe. The Groningen field was now evidence of deposits large enough to be commercially profitable. No drilling could take place until the Continental Shelf Convention, which set out which countries surrounding the North Sea owned which parts. An agreement had been struck in 1958, but Britain did not actually ratify it until 1964 when a meridian line was drawn between the UK and Norway. Yet if the officials had initially dragged their black brogue heels, once the pen had touched paper they switched to running spikes and were one of the first nations to issue licences for exploration and production.

Geologists had taken a new approach to thinking about the formation of rock layers, based on plate tectonics and what is known as continental drift which they believed led to the deposits of layers of porous rock, known as sedimentary basins, in the North Sea. They also believed that other geological events, in combination with pressures on the land, had caused faulting and, in turn, the creation of traps which might contain oil and gas. It was a theory they were keen to explore. Yet it was expensive. Offshore drilling was four times as expensive as on land and nobody had tackled it in such depths, though it was not new. In 1908 a primary well was sunk in the Persian Gulf, while marine drilling had already taken place in the coastal waters of the Caspian Sea and Lake Maracaibo in Venezuela, where equipment was carted out in barges and rigs erected from the seabed. In 1947 the first well was sunk out of sight of land in the Gulf of Mexico.

In September 1964 the British government divided up the North Sea into quadrants of 1 degree latitude and 1 degree longitude, with every quadrant subdivided into 30 blocks, each measuring 10 minutes of latitude and 12 minutes of longitude. In total 53 licences for the southern North Sea were issued to 22 consortiums, representing a total of 51 countries. The number of blocks issued was 394, and the first exploration well was sunk before the end of the year. Gas was the principal target, one struck by BP (formerly the Anglo-Iranian Company) in West Sole. When production began in 1967, this southern gas basin would begin converting the British nation, both domestic and industrial, from coal to gas.

The second licensing round took place in 1965 after Harold Wilson was elected in a landslide victory. The Labour government's main concern was to reduce the nation's balance of payments burden, and as the importation of oil was costing £300 million each year, Wilson was anxious to secure the nation's own supply. The next round involved 1,120 blocks, divided up into 127 areas and purchased by 44 different companies. Despite a small oil discovery in Danish waters in 1967, there was doubt among many oil men that large commercial oil deposits would be discovered. Yet two men – Edwin van der Bark of Philips Petroleum and George Williams of Shell – were convinced the oil would be found, but that it lay further north. The gas fields lay between the 53rd and 54th parallel; the pair believed oil would be found above the 56th parallel.

The first oil strike in the North Sea was on 23 June 1969. The Philips Petroleum company were preparing to pack up after digging 32 holes without success, but since their contractors were still on the clock they decided on one last attempt. They then struck what would become known as the Ekofisk oilfield containing 2.8 billion barrels, then one of the twenty largest fields in the world.

The *Aberdeen Press & Journal* would carry the news of the oil strike on the front page. In October 1970 BP finally raised up a core sample: 'a column of almost pure white sand, any trace of oil appeared to have been bleached out of it, but there was just a whiff of it'. Six months earlier Sir Eric Drake, BP's chairman, had said: 'There won't be a major field in the Northern Sea.' The discovery would become the Forties Field, which contained 1.8 billion barrels. The potential of North Sea oil was now apparent and the fourth round of licensing was scheduled to take place in August 1971, under the Conservative government who were now in power, albeit briefly. While Labour had maintained a tight control, the new Prime Minister, Edward Heath, reversed this. In a move that was later criticized by the parliamentary committee of public accounts, the remaining 278 blocks were put out to auction using a system of sealed bids, a method that would accrue the government just £41 million, almost half of which came from Shell and Esso's bid on what it believed to be 'the golden block' of the Brent field. In the audience at the London offices of the Department of Energy was a man on whom MI5 kept watch: Armand Hammer.

•

Occidental's headquarters during the company's bid to secure a prime slice of the North Sea was to be Claridges Hotel. While in London, the art deco jewel in the heart of Knightsbridge, once favoured by Winston Churchill, was now to be Dr Hammer's second home. It was here that, according to one former employee, he personally interviewed a string of beautiful escorts on their particular bedroom skills, before later introducing them to British officials whom he believed could assist his bid after the correct persuasion had been applied. While geologists and seismologists were dispatched to the chill of the North Sea, Dr Hammer toured restaurants and country homes to set up a consortium to fund the new venture. The first one to come on board was Jean Paul Getty, the billionaire oil baron who had exchanged the harsh light of California for the soft rain of Guildford and a sixteenth-century Tudor estate, Sutton Place, into which he had a payphone fitted. Getty already held stock in Occidental, but was always on the lookout for an acceptable risk that might bring a generous reward. Dr Hammer believed that the government would look favourably on any bid containing a

Scottish element, which might act as a sop to the growing clamour of nationalists, so he issued an invitation to Lord Thomson of Fleet, who was Canadian. The oil baron was to be joined by a press baron who, as well as owning a screed of Canadian newspapers, in addition to *The Times* of London and *The Sunday Times*, also owned the *Scotsman* and Scottish Television which he famously described as 'a licence to print money'. The offer appealed to Lord Thomson who, according to Hammer, said: 'I've always wanted to be a billionaire and never came close. God knows I'll never get there with my newspapers. But maybe this will give me my chance.'

It was a chance for which Thomson was reluctant to pay upfront, so a deal was struck that lowered his down payment on the promise that the sum would be reclaimed from future profits. (When Getty later tried to buy him out on the grounds that the wells might prove dry and a billionaire could better afford the loss, Thomson was only convinced of their ultimate success.) The 'Group', as the consortium was known, was to be led by Occidental, as operating partner, with 36.5 per cent, Thomson Enterprises, 20 per cent, Allied Chemicals, 20 per cent and J. Paul Getty, who took 23.5 per cent.

On 15 March 1972 the Group secured a North Sea petroleum production licence, an achievement celebrated with lunch at an expensive London restaurant. Yet, as befits a trio of men who never encountered a dollar they did not wish to keep, when confronted by the bill all were discovered to be wallet-less, each man having assumed one of the others would pay. This role fell to a passing millionaire.

Occidental's blocks lay in the northern North Sea, 130 miles east of the Orkney Islands to which the Ocean Victory – a self-propelled, semi-submersible drilling rig, which towered 29 stor-eys tall and cost $40,000 per day to operate – was dispatched. The geologists believed the most promising oil-bearing strata lay at a depth of between 8,000 and 12,000 feet. After ten months and two dry wells, the Ocean Victory struck oil in January 1973. The field was estimated to contain between 642 and 925 million barrels, of which 65 per cent were recoverable. Occidental christened the field Piper. In May 1974 Ocean Victory discovered a second field, 20 miles to the west of Piper, which contained 400 million barrels and which was called Claymore.

The discovery of the Piper field was crucial. Occidental's stock had collapsed in 1971, dropping 75 per cent due to the fall in tanker rates, while relations with Qaddafi were to prove difficult. The North Sea was not only an answer to their problems but promised to elevate the company's fortunes – Hammer hoped to the point when it could join the seven sisters as an eighth sibling. 'Wait until you see what the North Sea discovery does for this company,' he told Carl Blumay, his PR man. 'I'm a genius.' He was lucky too. The Yom Kippur war in October 1974 would trigger a major crash in the financial markets, but just before Egypt and Syria's tanks began crossing the Sinai desert, Occidental posted an advertisement to announce the financing for the development of the Piper field – $150 million for Occidental, $100 million for Thomson, with the finance underwritten by the Republican Bank of Dallas and the International Energy Bank. The company announced that output would be 220,000 barrels per day by July 1977, which at $10 would provide Occidental with an annual share of the gross revenue of around $67 million.

The focus now turned to constructing a platform and sourcing a suitable site at which to bring the oil and gas ashore. Lord Thomson, unwilling to invest in an extra salary, had given the job of running Thomson Scottish Petroleum to Alastair Dunnett, the editor of the *Scotsman* newspaper, who accepted the task in a spirit of stoicism mixed with nationalist pride and in the hope that what he lacked in salary today would be more than recouped in financial rewards tomorrow. It would not. Still, he took to the task vigorously, scouring the east coast of Scotland from the cockpit of a light aircraft for an appropriate landfall for Piper. The coast of Caithness was discounted as harsh and unwelcoming to ships, where 5,000 wrecks lie scattered in the sunken sand off the north of Scotland. The Orkney Islands proved more attractive, but two key sites came with different sets of problems. One was the principal nesting place for the world's population of long-tailed ducks. The second, Scapa Flow, was where the German High Seas Fleet scuttled their vessels in 1919; more importantly it was where the battleship *Royal Oak* was torpedoed in 1940 and sank with the loss of hundreds of lives. It was David Dunbar-Nasmyth, a retired admiral now assisting the Scottish Council to attract oil companies, who suggested the Orcadian island of Flotta to Dunnett as the solution. When in the Royal Navy he had taken a dreadnought into the surrounding waters, he explained on a flight the pair shared to Texas, which meant it was more than suitable for a tanker.

Armand Hammer would boast in his autobiography of winning over the residents of the island through charm and donating artworks to the museum in Kirkwall, but it was Dunnett who transformed himself into 'an Orcadian for several years' and persuaded those crofters whose land lay on the site of the prospective plant to sell. When it was finally completed the Flotta terminal stretched over 385 acres, covering a tenth of the island, and its dominant feature was the massive oil-storage tanks capable of containing 4.5 million barrels. It was unlikely to blend into the countryside, but this did not prevent a lengthy debate over which shade of paint would best act as camouflage.

While Alastair Dunnett was knocking on doors on Flotta, Armand Hammer was knocking together heads, or trying to, a situation which resulted in him having to reach for the aspirin.

The Piper platform, as he later wrote, 'had given me a brow-busting headache'. The cause of the problem was time. Hammer was late entering the cold waters of the North Sea, but this did not stop him wanting to be first to pump its riches ashore. The 'Monster', as Piper Alpha was nicknamed, was to be 495 feet tall and weigh 14,000 tons. It was to be assembled in Scotland at Ardersier shipyard by a subsidiary of the main contractor, J. Ray McDermott and Co of New Orleans. The work began in early autumn 1973, with Dr Hammer adamant that it would be on site by June 1974; however, this was scuppered by a dispute between the respective unions representing the engineers and the boiler-makers over the exclusive right to the welding work. A strike was called, with the boilermakers refusing to cross the picket line, which lasted a month, by which time the weather 'window' – the summer period when calm seas would allow erection – had closed. Hammer considered the unions 'suicidally obstructive' and had the next rig, Claymore, built in Cherbourg by the French firm Union Industrielle d'Enterprise (UIE).

The 'Monster' was finally moved out into the North Sea by massive barges in June 1975. The first oil raised in the North Sea and pumped into a tanker took place on 11 June 1975, when an elderly semi-submersible Transworld 58 at the Argyll Field secured its place in history, to Dr Hammer's bitter regret. The 'jacket' of Piper Alpha – the steel structure which sits on the seabed and rises up above the waves and onto which the platform is then attached – had a difficult birth. Despite favour-able weather reports, as the jacket reached the site a storm blew up, with winds of 90 miles an hour driving waves up to a height of 60 feet. It lasted two days. In order to save the jacket it was ballasted down and set upright on the sea floor until the foul weather passed, at which point it was refloated and then set down on its proper location come the command, in just 40 seconds.

As Dr Hammer wrote: 'But then it started to slip and, in the nick of time, pilings were hurriedly pounded down to secure it. The fact that the crew not only accomplished this feat, but also survived to tell about it paid tribute to its bravery, ingenuity and technological skill.'

The connection between Piper Alpha, its neighbouring platform the Claymore and the processing plant at Flotta was 128 miles of pipeline. The 30-inch pipe came in lengths of 39 feet, each costing around £1,000. To protect against the seawater each pipe was coated in a thick layer of anti-corrosive material, with a layer of cement added to prevent it from buckling. The pipe-laying process was performed by two barges; a trenching barge, which gouged out a shallow trench on the seabed, was then followed by a pipe-laying barge, which was flat, around 300 feet long with a railing round it, and which ran for 24 hours a day spinning out steel strands of pipe. On the barge the individual pipes were fed into a large tunnel and would pass through six different welding stations before moving down a curved gantry and into the water. It was now attached to the pipeline and would wait until enough subsequent pipes were attached to push it down through 400 feet of water to the waiting trench. Divers would then check that the welds were watertight before burying the pipeline. The largest barges could lay 1,700 metres of pipeline per day.

In truth Armand Hammer found the move into the North Sea difficult. It was, as he wrote:

embedded in a mare's nest of political complications and tangled in the cobwebs of age-old disputes between labour unions. In some ways, making the deal in the United Kingdom and getting the oil to flow combined the worst headaches of trying to do business in Russia and Libya. In Britain at that time the impediments of a heavy-handed bureaucracy were magnified by zealotry on the fringes of political power.

When Harold Wilson took power in 1974 after Edward Heath lost two elections, the threat of nationalization began to re-emerge, but it was never strong enough to worry Hammer extensively. He got on well with Tony Benn, Secretary of State for Energy, whom he regaled with tales of revolutionary Russia and his friendship with Lenin. He said Benn's hard bright eyes reminded him of Trotsky, although he was suspicious of his abstinence from alcohol and irritated by his perpetual pipe smoking.

•

Among the first Aberdonian employees of Occidental was Kate Graham, a 24-year-old who had recently returned from two years in Australia, where she had worked as an assessor for the Medical Benefits Fund in Sydney. In March 1975 an employment agency run by a friend sent her to interview for the post of secretary to Elmer Leon Daniel, a tall, thin Texan who was vice-president and operations manager for Occidental UK. An experienced driller, Leon Daniel (as he was known) had made the transition from the oilfields to the office without losing his dry sense of humour and laid-back manner. Kate's interview in the company's new offices at Causewayend, the area of Aberdeen where she was raised and went to school, consisted of little more than her pointing out the local landmarks.

The daughter of a tram-driver and a shop assistant, Graham was raised with a staunch work ethic and a curious nature that led her to investigate all aspects of the new world in which she found herself. The oil industry in Aberdeen was in its infancy and Occidental was initially little known, a fact that began to change when the opening of the Causewayend headquarters was covered by Grampian Television. While the first office dinner dance, which Graham organized, was for 120 people, sixty staff and spouses, the figure would grow to 700.

From a key position in Daniel's office, Graham completed her own curriculum on the oil industry, persistently interrogating staff on the meaning of each abbreviation and acronym. At her first board meeting she had to stop and check that the figures discussed around the table were actually in the 'millions'; with this confirmed, she returned to her typewriter and began 'gaily typing the zeros before the point'. She watched as the office began to fill up with Americans, laconic Texans and the 'Coon-asses' (as the oilmen of Louisiana were called), one of whom insisted on flying in crates of crayfish from his home state for the parties, of which there were plenty. 'There were stetsons and belt buckles and snakeskin boots,' said Graham. 'Dallas was on the go at the time and we had our JRs.' There was also a discreet band of staff united by the fact that each had been kicked out of Libya, one for failing to allow every employee to attend a speech by Colonel Gaddafi.

•

The tap was turned and the oil finally began to flow on its 128-mile journey from Piper Alpha to Flotta on 27 December 1976; a few weeks later than planned, as the Americans had wanted the oil to arrive on the anniversary of Pearl Harbor, 7 December, an inauspicious date which a few in the company were glad to have missed. On 11 January the airport at Kirkwall was stacked with private jets as the investors arrived at Flotta for the official opening and the offloading of the first shipment of oil from Piper Alpha, a ceremony presided over by Tony Benn, who, on account of the presence of thousands of gallons of flammable hydro-carbons, left his pipe in his pocket.

The discovery and construction of the Piper field was documented in a film shown in Britain and Europe, called *The Billion Dollar Bonanza*, one that Piper Alpha quickly lived up to when in 1979 the platform would produce 317,000 barrels of oil in one day, more than any single platform in the world, before or since.

3. ONE IN TEN THOUSAND YEARS

Robert Ballantyne was not unaware of the visit of Sue Jane Taylor. He had spotted the artist at work, sketching out on a gantry, but decided not to approach her on the grounds that she could be an 'Occy' spy. This was a pity, for if anyone on-board Piper Alpha was to appreciate Taylor's ambition to find beauty among barren mechanics it would be Bob Ballantyne. As a child he'd played truant to wander the streets of Glasgow, marvelling at the architecture of Charles Rennie Macintosh, and even now had in his cabin examples from the European Enlightenment. Ballantyne might have left school at fifteen without qualifications, but like many trade unionists his appetite for knowledge – first whetted by the works of Karl Marx and the speeches of John Maclean (the conscience of the Red Clydeside) – had expanded into art, history and culture. An electrician by trade, he had recently embarked on an Open University course on the Enlightenment and during his few hours off would study in an empty cabin – provided by the galley boss on condition that he didn't smoke or lie on the bed, and cleared up his own coffee cups.

He had a salt-and-pepper beard, grey-brown hair that kissed his collar and eyes that sparkled with mischief. His laughter was frequent and warm and he could deliver an anecdote like Billy Connelly. He was also lying low as far as his union activities were concerned.

Ballantyne had first arrived in the North Sea with electrical contractors James Scott in 1977, having previously worked at the Grangemouth petrochemical plant, followed by two years in the deserts of Libya where he beat the country's prohibition on alcohol by brewing his own beer and then trading it for grappa with the Italian workers. Each day he brought his trade unionist conscience to work with his sandwiches, and tried to improve conditions as well as recruit new members. In the North Sea this could be a dispiriting task. The workforce was divided amongst as many as thirty separate contractors with many men unwilling to risk a steady income on a wildcat strike for better hours or conditions. Others, meanwhile, welcomed the opportunity to work fifteen or eighteen hours daily despite the health risks and the precedence this set for more reluctant colleagues. Over the years Ballantyne had led sit-ins, battled with management and struggled to show potential members the benefits of union membership. The consequence of his political labour was to be persistently presented with an NRB (Not Required Back) certificate – fired, in effect, for agitation.

He had married at nineteen and been presented with a son shortly after his twentieth birthday; while his wife Eleanor, the daughter of a miner, was strongly supportive of his union work, cracks had appeared in the relationship that seemed to widen with each visit home. The adopted son of a joiner and small businessman father and a sickly, illiterate mother, Ballantyne had been scarred by his parents' own divorce and his sense of being unwanted by his birth mother. During his visits home he felt that his presence for a fortnight each month had become an intrusion into the smooth running of the household. Then there was the drink.

Like many oil men, he would race his colleagues to the pub or the off-licence to collect a carry-out for the train back to Glasgow and frequently be drunk by the time the carriage reached Stonehaven station, just 12 miles down the track.

Although he was able to put the cork in the bottle and, almost overnight, abstain completely, he was unwilling to remain in the marriage. While working in Stornoway at the Arnish Fabrication Yard, he had fallen in love with the daughter of the landlord at his guest house and so slipped away from his marriage, his son Gary and young daughter Amanda, believing it easier on Eleanor (and, perhaps, himself) to vanish than be a weekend dad assuaging guilt with treats.

Now, at the age of 45, Ballantyne enjoyed a degree of peace. He and Pat – a music graduate from Aberdeen University who was teaching – shared a small flat in Aberdeen and, thanks to favourable notices from a two-year contract in Holland, he was back offshore. He was first flown out to Piper Alpha in February 1987 as an electrician with Press Off-Shore, who had won the contract to complete a large refurbishment programme on the platform. When Press Off-Shore lost the contract to the Wood Group in December, Ballantyne, as was common practice, was simply rehired by the new firm. As his reputation for union action preceded him, when asked by fellow workers if he was 'hiding away', Ballantyne would reply that he was 'a wee bit fed up' losing his job. And now was not the time to lose work.

The current economic climate created foul weather in the North Sea. In 1981 the price per barrel had nudged $40, from where it steadily fell to $27 by 1984. A fresh oil crisis in the Middle East – where two of the world's largest oil producers, Iraq and Iran, were continuing to wage war on each other and funding their armies by overstepping their quotas – sent the price spiralling. When OPEC's control began to slip, the price per barrel fell from $30 in November 1985 to a paltry $10 in April 1986. The consequence in the North Sea was severe. Oil production fell from £20 billion to under £10 billion, drilling dropped by 40 per cent and prospective new fields were abandoned in favour of the cheaper option, which was to expand existing sites. Oil companies slashed budgets by 30–40 per cent while some contractors lengthened the working day from twelve to fifteen hours, with men no longer receiving pay for time off. In 1986 22,000 jobs were lost.

There were worse places than Piper Alpha to ride out a recession, or so Bob Ballantyne thought. The rig had a reputation. One bear said it would be 'the first rig on the moon', so explosive was its potential, but other platforms also had bad names and Ballantyne paid them little heed. Every platform on which he had worked was a petrochemical time bomb as far as he was concerned, and in his view Piper Alpha was better than others. But every platform had its disturbing tales. The riggers spoke of how the turbines produced gases that bubbled up to the surface of the sea and would ignite. The painters talked of being forbidden from using a grit gun to blast off the paint on specific ageing pipes that were so thin and corroded that there was the constant danger of a fracture. Instead they used a wire brush and boasted that 'if it wasn't for our paint these platforms would fall away into the sea'. While the scaffolders insisted the platform was sinking under the excess weight of each new module fitted, Piper's weight had risen to 34,000 tons. The spider deck was supposed to be 20 feet above sea level, but using the fixed length of scaffolding pipes as a ruler they figured it to be a few feet less.

The fact that Piper Alpha had an integrated health and safety committee that united the contractors was a step up from previous platforms on which Ballantyne had worked, although he was aware that as the personnel changed by rota, some participants considered it a way to skive off work for a few hours. Also, while the committee addressed the superficial issues such as ensuring that protective eyewear was worn, its power over more fundamental concerns, such as the positioning of the accommodation block directly above the processing modules, was negligible.

Like the rest of the workers on Piper Alpha, there was a side to the platform of which he was unaware.

•

On the helicopter flight out to Piper Alpha and her sister rigs, contractors were issued with a *Safety Handbook*, a small guide to health and safety which contained the stern declaration: 'study it well – it may be your passport to survival', but which would later be described as 'dangerously misleading' on account of faulty information. A method of throwing life-raft capsules over the side, which did not apply to Piper, was illustrated, with the instruction that they be entered via scramble nets, which the platform did not possess. (They had been removed three years previously.) Once on board, workers were given, at best, a perfunctory safety induction, or worse, none at all, with some entirely unaware of their own drill station.

In the North Sea safety had always taken a back seat to profit and the nation's energy requirements, and while Occidental was one of the few platforms to operate a safety committee, an incident in 1984 illustrated its limit. In March 1984, Piper Alpha suffered an explosion triggered by a gas escape in the gas conservation module. An attempt was made by a team of divers to put out the blaze using a fire hose, but they were driven back and instructions issued to abandon the platform. Winds of 30 knots had whipped up the sea, prohibiting evacuation by life-boat, so instead a Chinook helicopter with the seats ripped out for extra space was dispatched. Dozens of men were crammed on board. Three times the helicopter tried to land on the *Tharos*, a nearby vessel which was bobbing badly, and so the pilot had no choice but to head back to Aberdeen. Everyone was evacuated successfully, with only a few minor injuries, although one diver was almost blown off the helideck by the Chinook's down-draught and was saved by a colleague's quick reflexes and outstretched hand. On this occasion the *Tharos* also had problems extending its rescue ladder.

Although an internal board of inquiry conducted by Occidental investigated the explosion, the final report was restricted, with even the safety committee refused access. As a result the Occidental employee safety representatives resigned. (George Fowler, one of those who resigned in protest, would not survive the second explosion in 1988.) The Department of Energy, however, took a more lenient approach and decided not to prosecute Occidental on the grounds that the 'dangerous occurrence' was blamed on a 'design fault' which was 'probably attributable to the work of non-United Kingdom bodies'. The Minister was pacified by assurances that 'prompt corrective action had been taken by the operator'.

One of the consequences of the 1984 explosion was that a greater emphasis was placed on helicopters, rather than boats, as 'the favoured means of evacuation'. A memo prepared by Occidental's onshore Safety Superintendent was entitled 'How It Was vs How It Could Have Been'. The memo argued that the company had been extremely fortunate as the weather permitted helicopter landings on Piper Alpha and both containment and emergency support had been effective. Among the suggestions considered was to resite an alternative helideck, reposition life-boats and provide an in-field helicopter. It went on to raise potential problems, the key one being the ability of a vessel such as the *Tharos* to fight a prolonged fire that is being fed by a ruptured gas riser. Senior staff at Occidental felt the report was guilty of 'painting the worst case situation' and the recommendations were ignored. Piper Alpha, they insisted, was adequately prepared.

The threat of a prolonged high-pressure gas fire was a possibility but, according to Occidental, one so statistically improbable as to make the most diligent preventative measures not cost-effective. On 16 June 1987 a meeting took place at Occidental's headquarters in Aberdeen where senior management met to discuss a report compiled by a young engineer, Ian Saldana, which among other points detailed the consequence of a high-pressure gas fire. The report, which had been commissioned from the loss-prevention department in connection with their consideration of the need to continue with the hire of a Rapid Intervention Vehicle (RIV), described various scenarios that could weaken the platform's structural steel support members and how to fight against this. One possibility was an oil/gas riser rupture, the most serious consequence of which was a jet flame firing on supports. The author wrote: 'It is likely that an aerial deflagration from escaping gas or a fire on the sea from the escaping oil represents a more serious hazard to personnel and to platform abandonment plans than to the integrity of the structure itself and this may become the major concern in such an incident.'

Saldana was a Cassandra. Yet management's decision was to rely on the existing safety factors such as emergency safety valves on the risers. Subsea Isolation Valves (SSIVs) would allow the gas or oil to be shut off at various points along the seabed, yet the fitting of SSIVs would be expensive and necessitate that Piper Alpha be shut down while this was carried out. A report on 14 October 1986 by Elmslie Consultancy Services said: 'The lack of subsea valves on pipelines is an inherent hazard to the platform that it is impractical to resolve at this point of platform life.' In other words, Piper Alpha was considered to be too old, with the cost implications outweighing the paucity of years left in the platform's life.

Yet the Elmslie report had been clear on the result of a high-pressure gas fire:

These pipelines, especially the gas pipelines, would take hours to depressurise because of their capacity. This could result in a high-pressure gas fire on the cellar deck that would be virtually impossible to fight, and the protection systems would not be effective in providing the cooling needed for the duration of the depressurisation.

The possibility of such a gas-fuelled fire was considered to be slim. 'The probability of the event was so low that it was considered that it would not happen,' or 'It was not considered that in the lifetime of the platform there would be a situation where all the systems failed and that such a scenario would indeed occur,' a member of Occidental management afterwards explained.

This attitude was based on a calculation that such a catastrophic collapse of safety precautions was so remote as to occur once every 10,000 years, almost an infinity away for an industry present in the North Sea for less than twenty years. Yet this figure was a general assessment of the industry. No one sought to apply the same analysis to the specific platforms such as Piper or Claymore, a view that was later described by Lord Cullen as 'a dangerously superficial approach to a major hazard'. Yet the spectre of a high-pressure gas fire was constant. An Occidental memo stated that in the event of a fire fed by a large gas inventory the structural integrity of the platform could be lost within 15 minutes. It was dated 18 March 1988.

PART TWO

6 JULY 1988

4. THE LONG DAY'S JOURNEY INTO NIGHT

A seagull drifting on the light breeze that had blown up around dawn on Wednesday, 6 July 1988, would have looked down on a cluster of vessels gathered around the Piper Alpha platform whose occupants, like the birds above, were making the most of calm seas and clement weather.

The midsummer months were traditionally the period when crucial maintenance was carried out on the ageing structures in the North Sea, work whose precarious scaffolding and crews of painters, fitters and blasters could not cope with the wild storms and black crashing waves of winter. While Occidental carried out an extensive programme of repairs, maintenance and safety checks on Piper Alpha, the company had this year increased the burden on the platform by simultaneously embarking on a major construction project. A new pipeline riser was to be fitted onto the platform which would carry gas from the Chanter, a satellite field that had recently been tapped. A debate had taken place on whether, given the amount of work to be carried out, it would be safest to briefly stop production of oil and gas until the tasks were complete. However, it was decided that Occidental could not afford to allow the pipelines to run dry and so deprive the company – as well as the British government, which took 85 per cent – of almost $2 million per day in lost revenue.

Today, 6 July, which would be the platform's last, the production log would record an export of 138,294 barrels of oil, worth (at $14 a barrel) $1,936,116. Two of the vessels anchored in the blue waters just off the platform were there to assist in laying and fitting the Chanter pipeline and riser. The *Lowland Cavalier* was positioned 25 metres off the south-west corner of the platform with her stern facing Piper. The captain and crew were engaged in digging the trench in which the Coflexip pipe would subsequently be laid. A few days before, a survey technician, David Wiser (nicknamed 'Budweiser'), had transferred from the *Cavalier* to Piper to fix a beacon on to a small projecting deck and monitor its behaviour, a task that would take a few days and require him to stop over. Ed Punchard, the dive inspection controller, had been charged with showing him around, but not before introducing him to Stan MacLeod, who insisted he test his lung capacity on a polished brass object whose spinning wheel swiftly coated each 'victim' in a blizzard of talcum powder. 'Budweiser', who was 65, was old enough to take the joke, and began hanging out with the dive team when not at work.

The dive team on-board Piper Alpha had spent the past few weeks fitting a line of metal clamps to hold the new riser at depths of 50 and 120 feet, as well as a third clamp up on the spider deck. At deeper levels the clamps were put in position by saturation divers, based on-board the *Tharos*. Built in Japan at a cost of $100 million, Multi Support Vessel (MSV) *Tharos*, which weighed 30,000 tons, resembled a giant's coffee table on which sat a clutter of cranes, modules, accommodation block and helipad. A self-propelled, twin-hulled semi-submersible, she stood three storeys high, was supported by eight columns and rested on two pontoons the size of submarines. Designed with the support of Paul 'Red' Adair, the American oil well firefighter who inspired John Wayne's character in the film *Hellfighters*, the

Tharos, according to an article in *Oxy Today*, Occidental's in-house publication, was primarily designed to 'fight fires, kill oil wells and provide support and hospital facilities during any offshore emergency'. (Yet there had been accidents in the past. In 1982 a bridge connecting the *Tharos* to Piper Alpha collapsed with three fatalities.) She was currently anchored 550 metres off the west face of the platform. Like a pilotfish beside the bulk of a shark, the *Tharos* was accompanied by a smaller supply vessel, the *Maersk Cutter*, which was a mile away handling the vessel's anchors. The final link in the nautical chain that encircled the platform was the *Silver Pit*, a converted trawler; although on stand-by duty and required to be within five miles of the Piper, she was currently 400 metres off the north-west corner.

The previous weeks had not been without incident. On 3 June a plater using a blowtorch to mend a handrail situated 40 feet above the sea ignited a cloud of gas that had built up under the platform, having been discharged from a waste pipe. A blanket of fire 100 feet across burned for 30 minutes before a firefighting team extinguished it with a hosepipe. Twenty days later a mattress had caught fire in room C11 in the accommodation block – a steward who arrived to make the bed put it out, but the nearest fire detector failed to act. Afterwards when someone asked about the status lights that display the platform's state of alert, he was told there weren't any and to listen for the loudspeaker.

On 3 July, in order to accommodate work on the gas conservation module, Piper Alpha switched to what was known as Phase 1 gas production, meaning that all the gas recovered was no longer exported but was recycled or simply burned off. Meanwhile, oil production would continue as normal. Colin Seaton, the OIM, and his team were given no written instructions on how the plant should be run, on the grounds that they were sufficiently experienced; although the only time after 1979 (when gas exportation first began) when the platform had operated on Phase 1 production was for 60 days after the explosion in March 1984. The consequence in the switch to Phase 1 was that the volume of gas flared off each day rose from between 1–5 million standard cubic feet to as much as 30 million standard cubic feet, which resulted in an even fiercer heat from the flares. Workers' complaints of gas smells rose dramatically over the next couple of days, usually when one of the flare stacks would go out and there was a delay while a worker on a small protruding platform attempted to relight it by firing a flare-gun into the path of the gas cloud.

•

When Alex Rankin woke a little before 6 a.m. the morning of 6 July 1988, he rolled out of his narrow bunk bed on level B of the accommodation block and could be forgiven for thinking: 'just one more'.

As valve technician with Score (UK) Ltd, his company had been hired by Occidental to examine and recertify each of the approximately 300 pressure safety valves on Piper Alpha, a task begun in January with the arrival of a four-man crew who had progressed at a rate of three or four valves per day. The team had been demobilized on 11 April because the remaining valves would not be available until June and July when they returned, reduced to a two-man team for the final push towards completion. Rankin had arrived on-board Piper Alpha on 27 June, along with Terrance Sutton, a mechanical fitter, and both men had steadily worked through the remaining PSVs (as the pressure safety valves were known) until there was only one left to check.

This was PSV 504, which was located 15–20 feet above the floor of C Module where, to the naked eye, it disappeared amongst a maze of pipes, valves and cables. Its role was relatively simple; the valve would open automatically if the substance inside reached a preset pressure, allowing it safe means of escape and preventing a dangerous build-up. PSV 504 was fitted to Condensate Injection Pump 'A'.

There were two Condensate Injection Pumps and their role was crucial. Condensate is a form of liquid gas that is derived from the crude oil that rises on-board the production platform, the crude being run through a series of 'clean-up' processes designed to strip out unwanted elements such as water, sediment and wax; raw gas is also extracted and then flared off. A secondary element, a form of distillation, takes place whereby the lighter by-products of Liquefied Natural Gas (LNG) such as ethane, propane, methane and butane are extracted. These form a liquid called condensate. The condensate is then reinjected into the main oil line downstream of the main crude oil pumps, to begin its journey 120 miles to the Flotta terminal by means of three reciprocating compressors, giant gas turbines each weighing around 70 tons.

The Condensate Injection Pumps 'A' and 'B' were the only means by which the condensate could be pumped into the main oil line. One or the other or both ran 24 hours a day. Rankin hoped to get his hands on 'A' pump, finish the job and then go home. Before he could begin the task it was first required to prepare the necessary paperwork. The permit-to-work was a formal, written system to control specific tasks so as to ensure that potentially dangerous work was carried out with the appropriate safety measures and the knowledge of the relevant senior staff whose departments could be affected by the work in hand. On paper the system had appeared perfect. Before embarking on a task, a worker had to collect a form from the maintenance lead hand, fill in what he wished to do and the appropriate safety measures to be taken. The form had to be examined and then signed by the 'Designated Authority', usually the 'Control Room lead hand'. Then, at the end of the shift, the worker had to return the permits to the control room staff to sign off, stating either that the work was completed or was suspended until the following day. Under the system's guidelines it was mandatory that the worker and 'Designated Authority' met. In practice, they rarely did so because, as it was later found: 'the operating staff had no commitment to working to the written procedure; and that the procedure was knowingly and flagrantly disregarded.' The specific procedure was contained in Occidental's *Safety Procedures Manual*, yet staff were inadequately trained in its workings, the paperwork was at times poorly monitored, and the system itself regularly flouted. Contract staff found the permit-to-work system a nuisance and would often try to work round it. Multiple jobs were carried out on a single permit; forms that should be displayed were stuffed in the back pockets of an oily boiler suit, while others were left on desks unread. Although this was not yet apparent, it was disastrous.

The permit-to-work system had been put in the dock less than one year before, following the death of Frank Sutherland, a rigger employed by a contractor who died as a result of injuries sustained from a fall while attempting to attach heavy lifting gear to overhead beams on 7 September 1987. The only permit issued for the job read: 'to check and repair the thrust bearing'. There was no mention of lifting operations and no one applied for a fresh permit, which might have ensured the work was carried out. A few days after the death the legal department sent a memo to G.E. Grogan, the vice-president of engineering, which read:

I would confirm that there is significant exposure here and that prosecution is possible. I would therefore respectfully suggest that we proceed with care, particularly in our dealings with the Department of Energy. I would suggest that staff be reminded not to discuss the detail of the incident itself or follow-up investigation.

Occidental was prosecuted under the Health and Safety at Work Act and pleaded guilty to a complaint that specifically highlighted 'inadequate communication of information' between shifts. As a consequence, a memo was issued requesting more detail in job descriptions on the permits-to-work. Yet the results of the company's own board of inquiry report into the accident was not passed on to senior onshore personnel or even senior personnel on the Piper platform, who heard news of the conclusions 'on the grapevine'.

•

Andrew Rankin's exact introduction to the PTW system, like parts of his testimony, is confused. While management at Score later insisted that he and Terrance Sutton were briefed by the company's safety officer on its workings before departure, Rankin said he was not. He had just been promoted to supervisor before leaving for Piper, a position which made him responsible for PTWs. Although told that his training would be 'on the job', he had a rough knowledge of its workings from a previous visit, besides which there was a notice about the system pinned up on the wall of the Score container. So when he was asked, on arrival, by Occidental's maintenance superintendent if he knew the system, he said he was happy with it. The superintendent left it at that.

Shortly after Rankin had risen, showered and then dressed, he walked to the mezzanine level of D Module where he met with William H. Smith, the maintenance lead hand, who told him that 'A' pump had been shut down. There was work to be done on it and so PSV 504 would be available to him sometime during the day. In preparation, Rankin met up with Terrance

Sutton and together they strolled over to inspect the site which required both scaffolding to reach the valve and rigging to then lower it down. It was around 7 a.m., or shortly after, when Rankin picked up the PTW form and took it back to D Module where Kevan White, the maintenance supervisor, signed it, and also filled in parts which Rankin had missed out such as the tag number, PSV 504, and the location, C Module.

At about 7.40 a.m. Rankin took the PTW to the production office where it was signed by Bernard Curtis, the deputy production superintendent. A copy of this permit was later recovered from the sunken accommodation block. It was No. 23434 and was signed by both White and Curtis. On the permit under 'work to be done and equipment to be used' was written 'PSV refurbishment injection pump discharge condensate'. Under 'additional precautions' was written 'Open pipework to be fitted with blind flanges. Liaise with lead operator. Operator to isolate as required.'

It is at this point that the narrative differs depending on the person telling it. According to the later testimony of Andrew Rankin, he went straight to the control room, arrived at around 8 a.m. and spoke to a man he assumed was the lead operator, someone he could not later name but with whom he discussed the necessity of scaffolding and who then signed the PTW.

According to Joe Lynch, the lead production operator, Rankin spoke to him around 8.30 to 8.45 a.m. Lynch remembers Rankin saying he knew that 'A' pump was being handed to maintenance and so could he have access to PSV 504. It was a conversation witnessed by Harold Flook, one of the seven production operators on board the rig. Lynch said that at this point Rankin did not possess a PTW and so he sent him to William Smith to get one. (He was surprised to learn of the PTW signed by Curtis at 7.40 a.m.)

Regardless of whom Rankin met, he and Sutton then retired to the Score steel container, which was elevated above the floor at the far end of C Module, where they prepared their equipment and waited for 'A' pump to be isolated. Sutton had already spoken to a couple of riggers who said they would help them out, but it was not until after 2 p.m. that the scaffolding was up, complete with a green tag from the safety department.

Rankin then returned to the control room to retrieve the PTW. He stated that the form was filled in by a 'lead operator' whom he could not name, nor could he say if this was the same person he had dealt with on his first visit. He also said the 'lead operator' then telephoned Peter Grant, a phase 1 production operator, to check that the necessary electrical isolation of 'A' pump was complete. The visit lasted less than a minute, after which Rankin returned to the job site where he met with Sutton and also Peter Grant, to whom he showed the PTW form. Grant began checking that all the isolations were in fact complete. While these checks continued Rankin returned to the control room. He had been concerned about the absence of a rigger, but upon his return to C Module he discovered that John Rutherford had arrived and already rigged up a sling to take the weight of the pressure safety valve. This was then lowered first on to the scaffolding, then down 15–20 feet to the module floor, where James McDonald, a fellow rigger, trundled it in a wheelbarrow to the end of the module. There it was then lifted up by crane into the Score container. While Rankin began to check the PSV, his colleague Terrance Sutton returned to the location carrying a couple of steel discs the size of a dinner plate.

The blind (or blank) flanges were a means of protecting the highly polished flange seals that were open on both sides of the pipework now that the safety valve had been removed. They could be wrapped in duct tape or covered with a sheet of hardboard, but it was Score's policy to fit the blank flanges that were then secured with stud bolts. Standing on the scaffolding high above C Module, Terrance Sutton fitted the two flanges and screwed in the stud bolts as tightly as he could, using his fingers rather than a spanner as was common practice to prevent dirt entering the holes. While a blank flange could be used to seal a pipe, it would require to be checked, pressure-tested and signed off by production staff before use. This was not such an occasion. He reasoned that nothing else was required, as 'A' pump was electrically isolated and could not be used without the replacement of PSV 504. It took less than an hour for Sutton to complete the task and return to the container where he helped Rankin to finish up the various checks.

At 5.40 p.m. a completed test certificate was issued for the pump and witnessed by Neil McLeod, Occidental's Quality Assurance Inspector. The pump could have been refitted shortly afterwards, but word had drifted down that William Smith, the maintenance lead hand, had decided that the work would be completed the following day. This was not surprising, since it was common for Smith to terminate work sharply at 6 p.m. in order to avoid paying contractors' overtime rates. Yet once again there is confusion. While McDonald, the rigger, said he learned that the valve would stay in the Score container overnight from either Rutherford, his fellow rigger, or Terrance Sutton, Rankin said he was told this at 6 p.m. when he returned to the control room to arrange a crane to lift the valve back down on to the module floor.

According to Rankin, the 'oncoming lead operator' said no crane was available and so both men agreed the permission-to-work would be suspended. Rankin said that the operator retrieved the forms and that he, Rankin, then suspended it by writing 'SUSP' in the column on the form marked 'gas test'. This was the correct procedure, yet it cast doubt on the entire encounter as Rankin had never suspended a PTW before and later couldn't explain how he had managed to do it correctly. The form was either handed to the operator or left on his desk.

Back at the Score container, Rankin met Sutton and the two men knocked off work; although it was his responsibility to check that the site was safe prior to the suspension of work, Rankin did not return to 'A' pump. According to Rankin, after returning to their rooms to wash they both headed to the recreation area where they bumped into Smith. Rankin said that for his own peace of mind he wanted to tell him about the state of the PSV. Smith, he said, asked if the blind flanges had been fitted and Rankin said 'yes'.

The most likely explanation for what would subsequently occur was that Rankin, upon hearing that the job was to be postponed (possibly from Sutton, if he had indeed told the riggers) filled out the form in what he thought was the correct way to denote the task's suspension and then simply left it on the control room desk.

Between 5 p.m. and 6 p.m., a series of four changeovers took place as the day shift gave way to night and, in a combination of timing, systemic failure and human error, somehow PSV 504 was lost. While it remained disconnected and sat on the floor of the Score container, those in charge of Piper Alpha as the sky slowly darkened believed it was still in place, a steel sentry on duty.

The handover between the maintenance lead hand William Smith and his successor Alexander Clark took place in the maintenance office at around 5.30 p.m. This involved the two men going over a diary of the day's tasks, as well as an A4 pad on which changes were to be written. Unfortunately Smith made no mention of PSV 504, nor was it to be found in either diary or pad. He did, however, say that 'A' pump had been shut down and electrically isolated so that work could be performed on a Voith coupling, but that this work had not yet started.

The other three of the four changeovers took place among

the clean white floor and walls of dials and switches of the control room. The changeover between the phase 1 operators, from Peter Grant to Robert Richard, occurred at around 5.15 p.m., at the back of the panels, and out of sight of the other two switching teams. The basis of the changeover was to go through the phase 1 operator's logbook, in which Grant would be expected to record the removal of PSV 504 even if (as he may have done) he believed it to have been replaced again after its check had been completed. Yet, like Smith, he appears not to have alerted his successor to the work.

The control room operators' handover, those in overall charge of the room, began at 5.10 p.m. and lasted 5–10 minutes as Raymond Price brought his successor, Geoff Bollands, up to speed on the oil, water-injection and produced water plants as well as the diesel pumps and the JB turbines. The gas plant was not included in the handover as this was the prerogative of the phase 1 operators, and so there was no mention of PSV 504.

The changeover of greatest importance was that between the lead production operators, Harold Flook and Robert Vernon, the two men directly involved with processing the PTWs. Unfortunately the system on Piper Alpha had developed in such a way that at no point did switching staff go through the dozens of PTWs together. Instead the two men met at 5.10 p.m. and spent 20–25 minutes going through notes that were kept as an aide-memoire on an A4 pad. Flook should have told Vernon that PSV 504 was still off, as no paperwork confirming its completion had yet been received. Yet it appears he did not do so.

After the handover, at about 5.40 p.m., just as up in the Score container the completed test certificate was being signed off, Robert Vernon went off on the tour of the platform he completed each day. He returned to the control room at around 6 p.m., where he was expected to begin going through the accumulated

PTWs to check all was in order. It was routine for staff returning a PTW for completion or suspension and finding the lead operator unavailable, to then sign his copy of the form, match it up with the lead operator's two copies and leave them on the desk. And this is what Rankin appears to have done.

This piece of paper, specifically designed to convey crucial information, was lying on a desk and, for whatever reason, would go unread. It was as if it had now been tightly rolled and lit, like a fuse.

Meanwhile, tired after a long day, Andrew Rankin and Terrance Sutton decided to have dinner and then play a few frames of snooker. Rankin didn't fancy the night's film, *Caddyshack*, a comedy set on a golf course, and starring Bill Murray as a maniacal groundsman on the hunt for a tunnelling gopher. He decided instead to have an early night.

•

Carl Busse was a long way from home. The directional drilling supervisor on Piper Alpha was a quiet Texan from the small town of Navasota. The middle child of five siblings, Carl had grown up as the family's peacemaker, anxious to dampen down any argument he came across. Raised on the family's dairy farm, he was driving a tractor by the age of five, before he could even see over the steering wheel, and later helped his sister to raise calves. At the local high school, he joined the Future Farmers of America, played touch football and rodeoed, but like many young Texan men he had one eye on the oilfields and the other on a farm of his own. Busse figured one would pay for the other and after graduating from high school, he went to Houston and the offices of Eastman Woodstock – an international oil firm for whom a family friend, Rock Garrington, worked. Having got his foot on the bottom rung by talking the directors into letting him mow the grass on the office lot, he climbed up through the departments of inventory and purchasing, then surveying before going on to train as a directional driller.

At the age of 31, Carl had seen the world, having worked in Australia, Malaysia and Brunei as well as helping Red Adair to cap a blowout in Mexico. He was also already on his second marriage. He had married Kathy Martin in 1983 and their daughter, Fern, was born three years later in 1986, but they split up shortly afterwards and in March of this year (1988), Busse had tied the knot again, this time to Lou Brady, another girl from Navasota.

Busse had not been scheduled to undertake the drilling role on Piper, but a colleague who was due to fly out had suffered a death in the family and so he took his place. He didn't mind, having just returned from a holiday at home, water skiing, bailing hay and taking his daughter to Salem Lutheran Church, the small, white clapboard church where he always sat in the same place, second bench from the back on the right-hand side. In the past few weeks they had been deepening the wells, producing enough vibration to send the curtains in cabins swinging like pendulums. The first few days back had gone well and now that another one was over, Busse was looking forward to dinner and maybe a movie.

•

Down at the dive unit Stan MacLeod was in his office reading a copy of the *SAS Survival Handbook*, which he had been given as a fortieth birthday present. Punchard asked if he was considering a change of career, then settled down to plan the night's dives that he hoped might get as far as beginning the installation of the Chanter riser. The first dive, which began just after 6 p.m., involved Keith Cunningham making a final inspection on one of the clamps.

The dive team received a phone call at around 7 p.m. to say that a heavy load was about to be carried by the crane and so the divers had to step down until it was completed. When asked how long this would take, Punchard said it could be anything between one and three hours. A usual part of his daily routine was to head into the control room at 4.45 p.m. and collect a report that detailed how many of the fire pumps were in operation on any given night. However, he'd forgotten that day, and when he turned up a couple of hours later one of the men said: 'You're late.'

He then said: 'Nothing changes. It's exactly the same as usual.' The man then went to a filing cabinet, pulled out a blank form, checked the lines on the panel immediately behind the desk to confirm the status of the pumps and filled in the form. Two water pumps were switched to manual, which would allow them to be operated from the control room in an emergency, while the remainder of the pumps were switched off. This was in case a diver was sucked in through the intake pipe, which had previously happened to one diver, nicknamed 'Happy Day', who had been partially pulled in and injured his shoulder. Punchard now passed on the update to John Barr.

Shortly before 9 p.m., Punchard was approached by 'Lens'. Brian Lithgow, the photographic technician, who wore a tight pigtail and circular National Health glasses, was an eccentric character who drove a black London taxi as his personal car. 'Lens' was in the habit of phoning his girlfriend each night at around 9 p.m. and then, if work was quiet, he stayed for a few frames of snooker. When he asked if it was OK to head up to the accommodation block, Punchard nodded.

At 9.20 p.m. Gareth Parry-Davies was lowered into the water, armed with his grit gun.

•

At 9.45 p.m. the occupants of the control room were disturbed by a flashing light on the central panel which indicated that Condensate Injection Pump 'B' had switched off, or 'tripped' as was the term. As Bollands, the control-room operator, reached forward to push the button and acknowledge the alarm, Robert Vernon – who until a moment ago had been sitting beside him at the panels – was already out of the door and hurrying down to Module C where the pump was located. This was serious; if the condensate injection pump system failed, the whole rig could be forced to shut down, and no one wanted that to happen.

In the control room Bollands followed procedure and used the radio to contact Robert Richard, the phase 1 operator, who, once alerted to the situation, also headed for the scene. Then a second button lit up on the panel, followed by a beeping alarm. There was now a problem with the JT Flash Drum high-level alarm. Bollands pushed the 'accept' button, but it stayed on. He again called Richard, who understood that it was necessary to unload the reciprocating compressors because this would reduce the flow of condensate into the JT flash drum. Bollands was not yet panicked, as he had had similar experiences perhaps a dozen times before, yet there was a growing concern, for without the condensate pump the entire gas plant would shortly be lost and without the gas supply to the massive JB generators, they would shudder to a halt and the oil – or as Armand Hammer said on a visit to the platform, 'all that money being pumped ashore' – would cease.

There was also the spectre of a 'black start'. The generators were supposed to switch automatically to diesel in the event of a loss of gas, but if this failed, as sometimes happened, and the manual switch to diesel was unsuccessful, the drill could become stuck 9,000 feet down. Bollands figured they were now 30 minutes away from a shut-down.

Vernon returned to the control room a few minutes later and

told Bollands that 'B' pump would not restart. Bollands figured that hydrates were the problem, but Vernon said he had found a lot of lube oil around the pump. Vernon then said they should switch to 'A' pump, which he said was currently out for maintenance and was electrically isolated, but could be brought back on line. There was no mention of PSV 504. He then retrieved the PTW for 'A' pump from the box within the control room that held all the PTWs for the 68ft-level. The paperwork, once retrieved, had two red tags on it, which meant both the switchgear and the lube oil pump had been electrically isolated.

The PTW required two signatures and while Vernon supplied the first, Bollands called Alexander Clark (who was in the maintenance superintendent's office) on the tannoy to provide the second. When Clark called back, Bollands explained the problem. Clark agreed to the solution and, having tannoyed for an electrician to phone him back at the control room, he then headed directly there himself.

However, Vernon had already left the control room and returned to Module C when Clark arrived and found the paperwork already waiting. He was just about to sign when Savage, the electrician, phoned to say he was at the end of his shift. Clark, signing the papers as he spoke, told him not to bother, he would get one of the two night-shift electricians to do it.

It was around 9.57 p.m. when Vernon came down the stairs and back into C Module. He walked past the plate skimmer, the JT flash drum and the pump's main control panel. Then he passed Erland Grieve who asked: 'What's the score?' and told him he couldn't get the pumps to work. Vernon and Richards then made one last attempt to start 'B' pump. Richards went to the main control panel, JCP 057, to reset the system, while Vernon stood at the pump's push-pull button. Meanwhile, Grieve went to the local pump panel in readiness to push the 'start' button. Yet before they could begin Richards was called away. Vernon took over at the main control panel, but to little effect as the pump stubbornly refused to work. The electric motor kicked in, turned a few revolutions and then died.

Upstairs in the control room, the situation had deteriorated further as two centrifugal compressors tripped. Bollands estimated the plant was now 90 per cent into a shut-down situation.

Downstairs Vernon had turned his attention to 'A' pump. He would have reset the system on the main control panel, connected the airline and opened the push-pull button on the pump's gas-operated valves. In order to pressurize the discharge of 'A' pump, operators would perform a practice known as 'jagging' in which they repeatedly opened the gas-operated valve for short bursts. Vernon then 'jagged' once for a few seconds, then 'jagged' again for around 30 seconds.

In the control room a gas alarm light immediately came on, accompanied by a high-pitched alarm. Bollands snatched up the radio and called Richards, but found he couldn't hear him over the din, and anyway he now had to contend with a third centrifugal compressor tripping. Bollands was about to initiate a manual switchover of the main generators from fuel gas to diesel, so as to avert any loss of power, when four more gas alarms lit up. He tried once more to contact Richards by radio, but again the noise was deafening. Bollands was just reaching out his hand to silence the alarm when—

•

In Room B12 Harry Calder, the helicopter landing officer, turned off the light and pulled the covers tightly around him. In a separate cabin, Ian Fowler, a joiner with the Wood Group, was waiting for the racing results on the radio. Mark Reid, a lead foreman with Bawden drilling who had arrived on-board that morning and started work at around noon, was taking a break in the tea-shack. Down in the Bawden office John Gutteridge, the toolpusher, had switched on the radio for the 10 o'clock news; he was anxious to hear the latest about the Iranian passenger plane shot down by the Americans.

Meanwhile Bob Ballantyne, who had spent the day working in just a pair of overalls and his underpants because the heat from the east flare boom was so intense, was just leaving the television room. Having showered, dined, done some laundry and called Pat, he'd just heard that someone had phoned the radio room to get a copy of *The Whistle Blower*.

In the cinema Robert Pearston had joined dozens of others in the dark where they watched the movie *Caddyshack*, in which Rodney Dangerfield was just about to line up a shot he would not get the opportunity to complete.

On the *Tharos*, Charles Miller was heading outside with his camera. His children had asked him to get a photograph of the sunset for their school project.

The noise that preceded the explosion was described variously as 'a banshee's wail' and the death rattle of a 'strangled woman'. In the maintenance workshop Dave Ellington looked up, thinking it was the scaffolders 'acting the goat'. Ian Ferguson, who was working alongside him, put the blame on the air starter on a diver's unit.

In C Module Vernon had turned and was walking away from 'A' pump when the blast knocked him to the floor. Erland Grieve, also knocked down, rose up to see an orange ball of transparent flame, half the size of the pump skid, coming down through the roof between the two condensate injection pumps. He made his way towards the Ansul firefighting unit, but within 5–10 seconds the flame had gone out.

•

Bollands found himself hurtling 15 feet through the air before landing on his hip-bone, rising sore, shocked and with a deep gash on his left thumb. The room was in disarray, computers and telephones scattered, panels smashed and broken and a curtain of smoke rising from waist height. The air smelt of burning cable. He ducked under the smoke, reached the control panel and punched down the emergency shut-down button, then saw that Alex Clark was on his knees, badly dazed and bloodied, having been slammed face first into an oil-well production board. A few seconds later Ian 'Fergie' Ferguson, a mechanical technician, arrived in the room and helped Clark down the stairs. In a minute or two the three of them were outside, standing against the steel barriers at the north-west corner on the 84ft-level, panting deeply and trying to work out what the hell had happened. It was there that they noticed the silence. The rig had run down, its incessant din struck dumb, but where were the alarms? Robert Carroll, a safety operator who had joined them, began to smash the glass panels on the manual alarms, but nothing happened. The power was out.

Bollands knew that his own muster station was just 20 yards away, Boats 4 and 5, but the smoke had formed a 'black wall' that now ran across the front of them. Showers of sparks had begun to drop.

Soon after, the group was joined by Grieve, Young and Vernon. There was a discussion on the lack of water pressure and the need to switch the water pumps from manual back to automatic. At the time of the explosion, two of the diesel fire pumps which sucked in water for the fire hydrants had been switched to manual mode, while others were switched off altogether. It was routine practice during the summer diving season for the pumps to be kept on manual mode from 6 p.m. to 6 a.m. to prevent divers working in the vicinity being sucked into the intake pipe. If they were still operational, it would require manual intervention to restart the operation of the pumps, which could not be switched on again from the control room. The switch was on a panel 16 inches wide, located amid the smoke and fire of D Module. Vernon and Carroll insisted that they would go and do it. So after quickly suiting up in face masks and oxygen tanks, they headed off into the thickening smoke.

5. ANATOMY OF AN INFERNO

The cause of the initial ignition is unknown. It could have been as simple as a spanner lying on a rag.

What *is* known is that when Vernon pressed the button to start up injection pump 'A', he did so twice. The first short jag lasted only a second or two, but it was enough to force the condensate into the pipeline and up towards the blind flange fitted in place of Pressure Safety Valve 504 in C Module. The steel bolts on the flange, secured by hand, were not tight enough to hold back the pressurized condensate, which fanned out from the round steel disc like the sun's rays in a children's drawing. The first jag released a swirl of vapour, the equivalent of 4 kg per minute.

He then pressed the button again, this time for thirty seconds, which, at a rate of 110 kg per minute, allowed 45 kg of condensate to seep out and fill around 25 per cent of with a C Module flammable mixture. The escape triggered a flurry of gas alarms, but the mixture was already creeping towards its mysterious ignition point. The most likely explanation for the explosive trigger was an electrostatic spark. The release of liquid condensate under pressure produced a jet containing liquid droplets which were electrostatically charged. When these fell on an object, such as a spanner, which was in turn insulated from the earth (perhaps by means of a rag), an electrostatic charge built up. In seconds this discharged and ignited the whole cloud.

It was 10 p.m. and the only witness to the initial explosion was Michael Clegg, master of the *Lowland Cavalier*, a converted fishing vessel which was resting stern on, 25 metres off the platform. He described it as 'like the starting of a gas burner, a water heater. It seemed to go along the bottom of the platform; not out from the platform, but along within the platform, like a very light blue explosion.' The blue flame then retracted back inside the module. It would be another twenty seconds before the true consequence was felt.

This was the first in a chain of explosions which, like toppling dominoes, ignited one after the other and within just a few hours would lead to the total destruction of the rig.

Luck had turned her face away from the Piper Alpha and her crew that night, for had the initial explosion been able to vent freely through the east and west face of the module then this singular event might have resulted in only a handful of injured men. The problem was the environment of C Module. It was congested with equipment and pipework and so the deflagration generated enough pressure to destroy parts of the bolted latticework firewalls which acted as shields between the modules. The walls between B/C Modules and C/D Modules were designed to protect against fire and could hold an inferno at bay for between four and six hours, but in the face of an explosive blast they were useless and shattered instantly. The failure of the C/D firewall caused the loss of all firewater supplies and emergency power, as these facilities were vulnerable to damage in D Module.

As the steel walls collapsed, bolts and pieces of metal latticework were converted into projectiles which smashed into and ruptured pipework in B Module, next door, releasing a volatile crude oil which poured onto the floor and was soon ablaze. The fire quickly produced a dark, heavy plume of combustion products and unburned hot fuel-air mixture that was channelled by the structural beams and the prevailing wind through the platform to emerge from its east face where it billowed out over the sea and into the sky.

Back in B Module, a 4-inch-diameter pipeline carrying condensate from C Module to the Main Oil Line (MOL) in B Module had been damaged when the B/C firewall was destroyed. It failed suddenly as the Emergency Safety Valve (ESV) 208 was closing, releasing 112 kg of condensate into the flames in B Module. The result was a fireball bigger than a bungalow which burst out at least 20 metres from the west face of B Module. It was 33 metres tall, 23 metres wide and contained 112 kg of fuel. Charles Miller was on the deck of the *Tharos*, with his camera raised to capture Piper Alpha warmed in the glow of a summer sunset: a picture for his children's school project. The shutter clicked instead on a more immediate ball of burning gas. The fireball then decayed with every passing second.

B Module had been transformed into a blazing engine room that would drive the platform to destruction. The main source of fuel in the process plant was in the separators and was estimated at 50–55 tonnes. When this flooded out onto the floor of the module it formed a 'swimming' pool roughly 150 metres square to create a huge fire continually fed by oil leaking from the Main Oil Line through the Emergency Safety Valve 208, which failed to achieve a sufficiently tight shut-off to halt the flow of oil flooding back along the pipeline from the Claymore platform.

The explosion that gave birth to the fireball had a quieter but more deadly twin. While Charles Miller was dazzled and startled by the fire racing out from the rig, he was unaware of the puff of smoke and white light which suddenly appeared far from B Module, along Piper's north face. But his camera captured it. Unburnt, fuel-rich gases mixed from the fires in B Module and C Module had been swiftly pushed along the ceilings of corridors and stairwells until, high up on the 133ft-level at the base of the

Emergency Replacement Quarters, they had emerged into the air and ignited. The fire that began here would burn with a ferocity that reached 900 degrees, and produce clouds of smoke that infiltrated the accommodation quarters, choking and blinding the occupants.

The fire rose up and it also spread down. Oil leaking onto the deck of the 84ft-level had now started to spill over the collar of the deck penetration where the MOL rose up from the 68ft-level. Orange flames flowed down to the 68ft-level as burning condensate vapour was pushed through the hole in the floor around the Main Oil Line pipe penetration. Within a minute of the initial explosion, this was followed by burning crude oil, which ran down the MOL at the 68ft-level, creating a growing fire on the dive skid below. The flow of burning oil continued, and within five minutes the area beneath B Module was impassable.

Thus a second 'swimming' pool of burning oil was forming on the dive skid. This area was grated, which would normally have allowed the oil to flow down into the sea, but, according to later expert testimony, on this occasion the steel grates had been covered up with rubber matting by the divers in order to assist the smooth running of their operations. This allowed the oil to gather, pool and, in a few more minutes, combine with nearby drums of rigwash, which had been set alight by dripping, flaming oil, into a single inferno. The fire was indiscriminate in what it burned, but there was one particularly vulnerable piece of pipework.

The Tartan riser was a pipeline, 11½ miles long and 18 inches wide, which carried gas at high pressure along the seabed from the nearby Tartan A platform to Piper, which then piped it, via MCP-01, ashore. Upon reaching Piper, the pipeline snaked up from the seabed, between legs B5 and B4, and as it reached the 68ft-level it bent sharply from vertical to horizontal, then continued its journey southwards, suspended by three pipehangers. When it reached legs B3 and B2, it entered a right-angle bend heading east, rose vertically and moved into the 68ft-level deck plating and connected into the main pipeline shutdown valve, ESV 6, where shortly after 10.01 p.m., an uncontrolled fire now brightly burned.

The pipeline between Tartan A and Piper contained 18 million metres of standard cubic feet (MMSCF) of gas (450 tonnes), encased in a steel tube one inch thick.

If exposed to a temperature between 580 and 700 degrees Celsius, the strength of steel would weaken and be unable to counteract the internal pressure of the gas, which was held at 1,800 pounds per square inch (p.s.i) and so would eventually blow. If the pipe had been fireproofed it might have lasted one, two or maybe even three hours. But it was not. If the deluge system had been in operation the pipe would have been drenched in water, which would have helped to keep it cool. It was not, for it had been disabled in the initial explosion. (The deluge system in the gas compression module, had it worked, would have been restricted, since 50 per cent of it was blocked by scale, a fact discovered during a routine test in May but which had not yet been rectified.)

So the fire began to heat the pipeline and the steel temperature started to climb. Later tests would predict a similar pipe to fail within 7–18 minutes. The Tartan riser held out for 20 minutes.

Two minutes after the condensate was driven up and wriggled out from behind the steel bolts of the blind flange (where PSV 504 had once been) Piper Alpha was now primed, like a bomb, to blow. No one was aware of it, but for each man on board a clock had begun to tick. Within two hours – the length of a typical Hollywood movie – Piper Alpha would become a blackened skeleton of twisted steel. The accommodation blocks would be sunken coffins.

It happened as follows: At 10.20 p.m. the Tartan riser pipe was breached and the flames were immediately fed by 11 miles of pressurized gas, triggering a second massive explosion. A fireball blossomed out from below the centre of the jacket, enveloped the platform and rose to a height of 700 feet. Within sixty seconds 1, 260,000 cubic feet of gas had escaped and ignited, raising the temperature in the surrounding area by hundreds of degrees.

Like another domino toppling onto its partner, the destruction of the Tartan riser then put extra pressure and heat on a second gas line, also located in the same area. This was the MCP-01 gas line, which ran for 33½ miles from Piper Alpha to the MCP-01 processing platform and which contained 51 million standard cubic feet (MMSCF), the equivalent of 1,280 tonnes of gas. At 10.50 p.m., when it blew, the force was felt one mile away, debris was ejected over 800 metres and a fireball described as 'an inverted mushroom' flourished from beneath the rig and was projected 50 metres.

The platform could not withstand the heat generated by the two massive gas-fuelled fires which were now acting like blowtorches on support beams, floors and girders which began to buckle and bend. At 11.15 p.m. the western crane collapsed from its turret, sending the jib and cab falling into the sea. A few minutes later a major structural collapse at the centre of the platform began. The area around B Module – where the fires raged – faltered, which caused the drilling derrick to collapse towards the north-west corner with the top section falling across the pipe deck.

At 11.18 p.m. the Claymore gas riser – 16 inches wide, 21½ miles long and containing 10 MMSCF or 260 tonnes of gas – failed, resulting in yet another massive explosion, described by the master of the *Tharos* as 'the biggest of the night'. This explosion, combined with the accumulative effects of the previous explosions, was to lead to the further collapse of the structure. The pipe deck collapsed to the west at an angle of 45 degrees and was split from east to west along the line of the south face of the SPEE module. The 'White House' and OPG (offshore projects group) workshop, both of which were on the pipe deck, also collapsed.

The failure of the girders and deck which held up the East Replacement Quarters (ERQ) caused it to tip to the west and land on top of the Living Quarters West, first crushing it and then toppling it 130 feet into the sea. The ERQ then followed, both buildings sinking 430 feet to the sandy seabed below.

While in place both buildings had acted as a firewall, absorbing the heat and flame and partially protecting the Additional Accommodation West, which tipped in to fill the void left by the LQW and so was now subjected to extensive fire attack. It did not tip into the sea until 12.45 a.m.

The AAW was later discovered to have suffered much more extensive fire attack than the ERQ, with heating predominantly from the south. It tipped over on its north face and remained there for a period of time, though it could not have done so until the LQW had gone. Eyewitness accounts indicate that the north end of the platform collapsed slowly. Once the centre had fallen out of the platform the AAW would have been subjected to extreme heating on its south face, certainly from around 11.30 p.m.

By 12.15 a.m. the fire was mainly from the surface of the sea and was highly pressurized, like a Bunsen burner. It was fuelled by the burning gas coming up under pressure from one or all three of the gas risers, which by then had been severed by falling equipment and structural debris raining down from the north end of the platform.

In the early days of the Texas oilfields, a group of Pentecostal ministers visited the owner of a prominent oil company to raise their concerns. The extraction of so much oil had led them to worry that the fires of Hell might soon be extinguished. Had these men of the cloth been present on this night, their fears would have been assuaged.

6. 'GETTING BAD'

'Mayday. Mayday . . . explosion and fire on the oil rig on the platform and we'll [sic] abandoning abandoning the rig.'

The first message from David Kinrade, the radio operator, was broadcast on 2182 kHz, the emergency distress frequency, at 4½ minutes past ten. It conveyed the rising panic that had already begun to grip key personnel. The radio room was in a white portacabin at the top of the rig, and on one side overlooked the green baize of the helideck with its orange concentric circles reducing to the white 'H' at the centre. The view from the window on the other side was usually the mechanical clutter of the pipe deck, but smoke and flame now rose from it. The VHF radio sets had broken off the bulkheads; the TV set which had sat on top of the filing cabinet lay smashed on the floor. After the blast Kinrade picked up one of the multiple telephones, but there was no dial tone. The tannoy system was a small box with an attached microphone, but as he was still confused from the blast he was unsure whether to touch it in case it set off the general alarm, which required a specific order from the Offshore Installation Manager.

After a minute or so one of the other phones began to ring. Kinrade picked it up and heard the distressed voice of Eric Duncan, who was trapped several floors below in the materials office by a fire that had ignited outside the door. He wanted

John Dawson, the Telecoms engineer who was in the room with Kinrade, to come down with the key to the Telecoms room which was connected (via the production stores) to his room, and through which he might then escape. Kinrade said he would see what he could do.

Colin Seaton, the Offshore Installation manager, then arrived dressed in a survival suit and deeply agitated. Under the response outlined in Occidental's *Emergency Procedures Manual*, he would be expected to instruct Kinrade to establish a telephone link to the Occidental office in Aberdeen and contact both the standby vessel and the *Tharos* by VHF radio, as well as use the public address system to issue whatever instructions he deemed suitable to the men on board. Instead Seaton simply ordered him to put out a Mayday call and promptly left without issuing further instructions. A few seconds later he returned 'in a state of panic' and told them that the access way was on fire and full of smoke. Kinrade said they must get out, and that there was an escape hatch they could use. He then broke open a box and handed life jackets to Seaton and Dawson before putting on his own.

Seaton then told Kinrade to issue a message to abandon the rig, yet as the radio operator was also beginning to panic the message was haphazard. Before commencing he set off the two-tone alarm signal to draw attention to the message and warn off anyone else using the frequency.

At 10.06: 'Mayday Mayday . . . we require any assistance available any assistance available we've had an explosion and er . . . a very bad explosion and fire er . . . the Radio Room is badly damaged.'

At 10.08: 'Mayday Mayday . . . we're abandoning the Radio Room we're abandoning the Radio Room we can't talk any more we're on fire.'

Unfortunately, these messages were not audible on the

platform and Kinrade was not sure if the public address tannoy was working. The men then escaped through the hatch and made their way down to the galley.

•

The ferocity of the initial explosion rendered the majority of the actions listed in Occidental's *Emergency Procedures Manual* either impossible to perform or obsolete. It was as if the manual had imagined only those scenarios in which the heat is slowly turned up, permitting personnel to remain two or three steps ahead of a situation that escalates at a more leisurely pace.

If a fire broke out, or an explosion took place, the men nearest the scene were to activate the general emergency alarm or telephone the control room or radio room, whose operators would then alert the OIM, with personnel sent to investigate and report back. The OIM would immediately head to the radio room and from there direct the response. He would contact the Duty Communications Operator at the Occidental Emergency Control Centre in Aberdeen, liaise with the standby vessel, alert neighbouring installations, shipping and helicopters. He was also responsible for ordering the shutdown of processing and drilling, the direction of the firefighting crews and the evacuation of non-essential personnel.

This was to be organized by the Emergency Evacuation Controller (EEC) and his team, which consisted of the helicopter landing officer and lifeboat coxswains. The EEC and helicopter landing officer would meet in the reception area and ensure that the helideck was operational. Since helicopters were now the preferred mode of evacuation, personnel were drilled to assemble at their designated lifeboat muster station, from where their coxswains would lead them, one lifeboat compliment at a time, up to the helideck for evacuation. The lifeboats were to be lowered only as a last resort, and once the OIM had issued this specific instruction three times over the public address system.

The OIM was to be assisted in his tasks by the Operations Superintendent who was to head directly to the control room, assess the extent of the emergency, prioritize the response and coordinate the plant shutdown. He was also to deal directly with the three emergency teams who formed immediately after the first alarm sounded, each one consisting of six men and charged with the responsibility of remaining on the rig to fight any fires, retreating only when there was no longer any hope of exerting control. To ensure the plentiful supply of water for firefighting, a mechanic and electrician were to proceed to the electrical workshop, to start up and then run the emergency diesel pumps and the SOLAR generator.

There was no time for such a structured response. Michael Jennings, the Flight Information Logistics Officer (FILO), was responsible for coordinating the helicopter shuttles. He was sitting on the floor at the back of the small cinema when the floor jolted and panels and light fittings fell from the ceiling. The screen collapsed and for a few seconds the film was projected onto the bare wall before dissolving as the power to the projector died. 'It was as if you were in a car crash and you came to a sudden halt,' said Joseph Meanen, a scaffolder. The cinema was full that night with men, mostly dressed in jeans and jumpers. Though startled, they were reasonably calm as they began to crowd out into the corridors, confused by the lack of any ringing alarm. Jennings wasn't too bothered; he figured the gas con module had blown up again and that, as in 1984, there would be an orderly evacuation. He squeezed through the crowding men and headed upstairs to level 4 and the dining room and galley, where he picked up a phone and rang the radio room. John Dawson, the Telecom engineer, answered and said Kinrade had already put out a Mayday and that he should check out the stand-by radio room for organizing flights. The stand-by radio room was over on the west side of the platform, where there was a teleprinter and a telephone linked to a satellite. Jennings knew it would be a hassle to reach.

In the reception area, it became clearer still to Jennings that things were not as they should be. There were already two men, John Hackett and Jim Savage, equipped with breathing apparatus. A trickle of smoke had began to perfume the air and thickened each time someone opened (then quickly closed) the door that led out on to the deck. Andrew Mochan, a superintendent engineer with the Wood Group, had the role of redirecting people who were unable to access their lifeboat muster point. He had rushed up from his office, one floor below, to the reception area where he met Jim Heggie, a projects services superintendent who was also the Emergency Evacuation Controller. It quickly became apparent to both men that nobody was able to access their lifeboat muster point, and that it was essential to quickly identify a safe means of escape. Neil McLeod and Tommy Hayes then volunteered to put on breathing apparatus and head outside to try to find a way out. Just as the pair were pulling on their equipment, Alastair McDonald, a young technician in his early thirties, arrived at reception with burns to his face. In the toilet nearby he winced in pain and cried out as he tried to cool his burnt skin by throwing water over his face.

The corridors were hot and smoky. Jennings kept finding himself blocked off by superheated doors as he tried to get up to the radio room. It was clear that a great deal of heat was coming from the south side. After a detour down to his cabin to collect his life jacket, he braced himself and succeeded in getting out on to the covered stairwell leading up to the radio room; with each step he could feel the heat through his boots. The radio room had already been abandoned by the time he arrived, and looking out of the window Jennings knew that a helicopter evacuation was impossible: 'The smoke was very thick and black and was drifting right across the helideck. There were two helidecks there and it was drifting across both of them.' He stopped just long enough to collect a white hard hat and survival suit before turning round and heading back towards the reception area on level 4.

Meanwhile, in one cabin a man was sorting out his socks into pairs, rolling them together and putting them back in the drawer – to the incredulity of his room-mate, Bob Ballantyne. Ballantyne had been in the recreation room when the blast struck and immediately made for the cabin, where he found Ian Gillanders slightly dazed after the shower ceiling had collapsed on his head. A minute or two later his other room-mate, Charlie McLaughlin, also arrived. A bluff Glaswegian electrician, McLaughlin was fond of a drink and a regular tangle with management, but staunch in defending Catholic doctrine against attacks launched by Ballantyne, an atheist whom he dismissed good-naturedly as a 'blasphemer'. He was also first to organize a whip-round for charity, but last to pick up a towel from the floor. Gillanders, an instrument pipefitter, was as tidy and quiet as McLaughlin was loud, but together the three men got along very well. While Gillanders dressed, McLaughlin was fretting about whether between them they would have enough money for drink when they arrived back at the beach or perhaps even Bergen. They were gathering their wallets and passports and at first Ballantyne said he only had a few quid, to which McLaughlin replied that he would have to hook up with someone else. Ballantyne then assured him he had a cheque from his wife Pat. Then he went back upstairs to the canteen to see what was happening and realized the gravity of the situation as men hurried about in survival suits. A man he didn't recognize asked his name, cabin number and lifeboat station. After answering he returned to the cabin to discover that Gillanders (now dressed) was sorting out the laundry bag that Ballantyne had collected earlier in the evening.

'No, no, Ian, don't do this. Don't worry about it, we're leaving it, we're getting off the platform.'

To Ballantyne, it was as if Gillanders was trying to re-order the mess into which they had all fallen, one sock at a time. He told him to hurry up and then stopped to collect a few items of his own, his reading-glasses and the latest tome on his reading list of eighteenth-century literature, a copy of Voltaire's *Candide*, figuring it could be a long boat-ride back home. He stuffed the book into his survival suit. Each man's reaction was slightly different. While Ian appeared to be displaying symptoms of mild shock and an unwillingness to confront the grim reality of the situation, and Charlie was dampening any worries down under a blanket of humour, Bob – having seen the growing chaos upstairs – was becoming increasingly scared. Smoke was now penetrating the cabin and so he grabbed hand-towels, threw them in the sink, turned on the taps and then handed them out to his friends. Each man then wrapped one around his face and together they opened the cabin door. The corridor was crowded with men. At first they were at a loss as to what to do – which way to go? Ballantyne figured there was no point in heading back up to the galley with the crowds. After a quick discussion they decided to start checking the rooms in case anyone had been injured by the initial blast. It was also agreed that they would stick together.

Down on the drill floor a group of workers were securing the drilling operation. As soon as driller Fred Busby and toolpusher John Gutteridge had picked themselves up off the floor of the Bawden office now strewn with paper, books, broken chairs and smashed files (on level two of Additional Accommodation East) they had hauled on their boots and overalls and made their way down to the drill floor. Busby set off first and found that, due to the proximity of large flames licking over the handrails, he had to take a circuitous route that involved crossing the pipe deck walkway, which was just two feet wide and obscured by smoke. He held his breath, bent down and rushed across. Once on the drill floor he met up with five of his colleagues, Jeff Jones, James Gordon, Paul Williamson, Steven Rae and Vincent Swales.

Swales, the derrickman, was from Blairgowrie, and he had already tried to trigger an emergency alarm shortly after the explosion. He had smashed the glass covering with a wrench, but the glass merely folded in on itself 'like a jigsaw that would not come apart'. He was furious, as he had twice complained in recent weeks that the glass appeared faulty on this particular alarm, but the encroaching fire forced him to focus his mind on escape.

He had tried to go straight to his designated lifeboat, but had been forced back into the sack storage module by a curtain of flame, where he had met the mud engineer Ernie Gibson and three others. While Gibson wanted to try fighting the fire, Swales realized it was beyond control and led them by a circuitous route to the rig floor.

Swales helped Busby to secure the drill rig. The quickest way was to space the pipes in such a position that the pipe rams closed around them and closed the kelly cock (a flapper-type valve used to secure and close the drill). Busby then told everyone to head to the lifeboats; but as he was informed that this was impossible, he figured the best bet was to go back into the accommodation and up to the galley in preparation for a helicopter evacuation. Gutteridge organized a head count before they departed and discovered four Bawden men were missing, last seen in either the machine room or the engine room. The plan was to get to the accommodation unit and then track them down. Flames were rising 8–10 feet on either side of the rig, and the pipe deck which led to the accommodation block was a deep fog-bank of smoke.

'We all went across, or those we could see, with our hands on each other's shoulders so that we would not lose each other,' said Swales. 'And the smoke was coming across quite thick. We could see flames now and then at this level. There were red embers billowing over in the smoke, and we tried to dodge them as well.' The proximity of the fire and the inability to see clearly through the smoke agitated them, quickening their breathing. The more smoke they swallowed, the more they coughed and the tighter they tucked down their heads, forcing their chins onto their chests. Although linked together in a human chain, each man was alone in trying to hold his emotions together and push the fear to the back of his mind.

Before reaching the accommodation block the group passed through the electrical workshop, which Swales was reluctant to enter:

> It looked as though somebody had completely put it through 360 degrees and put it back; there was nothing on any shelves or on the work bench. The door was completely open; I can't remember if it was on or off. When I looked in the offices you could see the smoke down from the ceiling, the false ceiling was hanging down as well.

He lingered outside the door while the others passed through; a jittery, excitable character who knew his own mind, he decided at the last minute not to follow them and instead struck out on his own, heading east and feeling along the outside wall of the offices as he went.

•

Back inside the accommodation block, Busby found torches and told the crew to head up to the galley, while he and Gutteridge stopped by the Bawden office where they met Leslie Morris, the platform superintendent, who had been off-duty but was now anxious to help. He told them: 'There's been an explosion, but we need to find out what we need to do to put it right.'

Working separately, each man began to check every bedroom door as they progressed up the different levels, in case anyone was still asleep. Each exit from the accommodation block confronted them with flames or intense heat. Turning one corner, Busby and Gutteridge met a man in breathing apparatus: Alan Wicks, the Safety Supervisor. Busby told him about the men trapped in the mechanics' workshop, but Wicks was at a loss as to what he could do so Busby suggested he check out the westside accommodation for any exits. Gutteridge, who was in charge of a fire team that was supposed to muster at the White House, took up Busby's offer of a breathing set and hung on to him as he made his way to the west door on C deck level. However, the kit which was supposed to be there was already gone and Gutteridge was forced back choking and spluttering. His last image of Wicks was disappearing into the smoke; he never saw him again. Gutteridge and Busby then went to the nearest bedroom, collected towels in the bathroom, drenched them in the basin and, covering their faces, headed up to the galley.

They were confronted by a scene of confusion and rising panic. The long dining room had a dozen rectangular tables, plastic and metal chairs, and a few large potted plants and tartan curtains to lend it a more homely look. In the past 15 minutes as many as 100 men had gathered here, all looking for a secure exit. Yet as each door to the outside was opened and then closed on fire and flame, more smoke had crept in. When someone opened the emergency exit 'it was bright orange flames ... the black smoke pouring'. One group had unrolled the fire hose in an attempt to douse the door, only to find it pressureless and dry. Ian Fowler, a joiner, tried four times to escape out of one door that led to a landing on the south side of the galley, but each time was confronted by 'thick black smoke, sparks, a tremendous amount of heat'. The explosion had buckled doors on lower levels through which fire and smoke now penetrated. In a nearby stairwell the platform's medic was assisting an injured man whose face was burned and who appeared to be in deep shock; his face, in spite of the rawness and blistered broken skin, was blank and expressionless.

Those members of the emergency evacuation teams equipped with hand-held radios struggled to contact their colleagues who were wandering down dark corridors in search of a viable way out. When Michael Jennings returned to the reception area, where he put on his survival suit, he heard Jim Heggie, an evacuation controller, talking over his hand-held radio to Robbie Carroll, a safety officer, and knew by his response that the situation was 'pretty grim'. When Hayes and MacLeod returned – sweaty, hair and faces blackened with smoke and with their oxygen tanks drained – their report was similarly depressing. When another fire team gave the thumbs-up to a scaffolder who was peering through a window from the galley kitchen as they picked their way through the smoke, it was little more than a vain gesture of hope.

When yet more men poured into the reception area, a crush developed as they shoved each other into the galley. There was shouting, pushing and calls of 'What the fuck is going on?' Michael Jennings was carried along by the crowd, and startled to see through the windows the flames rising up the north face. Since the explosion the room had been poorly lit by strips of emergency lights, but around 10.18 p.m. these flickered out. The onset of darkness triggered a cacophony of calls and cries from the crowded men who began to spin around to see if another area offered more light, banging into chairs and tables while they moved. When those members of management who were present switched on their torches, men flocked around them and began to hurl abuse and demands for a swift solution. With arms outstretched, managers appealed for calm while struggling to contain their own fears.

Two minutes later the entire room was lit up. The Tartan riser had blown.

•

The ballooning canopy of fire and gas forced a number of men to commit themselves to a fall which, until now, they had kept putting off. After his reluctance to follow Busby, Gutteridge and the Bawden boys into the electrical workshop and on into the accommodation block, Vincent Swales had moved east along the outside of the office, bumping into a man in breathing apparatus; it was Alan Wicks, and together they reached the Aqua-Chem (a piece of machinery desalinating water for use on the platform) where they met two other men with BA kits. As Swales was unassisted by an oxygen tank, he was unable to follow them and so turned back. Shortly afterwards he met Steven Rae, an electrician also with Bawden Drilling who also had heeded his gut instinct and decided not to re-enter the accommodation block. Together they decided to head back to the rig floor. As Swales said: 'At least we can get a breath of air before we start again.'

The route back was smothered in smoke and Rae was only able to follow Swales by keeping focus on the reflective strip on his overalls. Once up on the rig floor they were distracted by the ringing of a telephone which Swales answered. It was Sid McBoyle, the night mechanic, trapped with a few other men in the mechanical workshop and anxious for news of a clear route out. Swales could only explain the route he had been taking, but while on the phone he could see the flames moving closer to the men's location and urged Sid not to stay. Both men hung up. Swales and Rae each collected life jackets from an office and carried them down to the skid deck, their goal being to reach the 20ft-level and escape into the water. Passing the well-heads, Swales saw the extent of the destruction:

> There is a large hatch for each single well, twelve across in rows of three. The small hatches in the middle of the larger hatches had a lot of them blown off. Some of the larger hatches were blown off and leaning to the side at an angle. You could see through into the production deck area and there were fires down there.

The sight of the fires curdled their stomachs, quickened their steps and brought beads of sweat to their brows. Swales became increasingly animated, running back and forth in search of the most suitable route down to sea level.

The pair were blocked or driven back along a number of routes and eventually found themselves on the walkway at the south-east corner. They were safe from smoke for now, but trapped by flames to the west and impenetrable smoke to the east. Swales was unable to stay still and kept moving back and forth along the walkway, repeatedly climbing over the handrail looking for any way down that did not involve a long leap. They were 68 feet above the water and, unable to find a rope, it quickly became apparent that they would have no option but to jump. But not just yet. Both agreed they would wait until they absolutely had to do so.

They took off their boots and put on their life jackets. Peering down into the water, Swales was worried at what he thought was oil on the surface directly beneath them, and told Rae he didn't want to have to leap into a burning sea should the debris that was flying past ignite it. Leaning out from the railing, he could see down to the north-west corner where a zodiac boat was coming in to collect men by the boat buffer, and so he began to wave.

The explosion prompted him to jump and Rae to follow. Remembering his survival school training, Swales held his nose with one hand and used the other to cross his chest and hold down the life jacket. Although he jumped feet first, he went into a somersault before hitting the surface and sank deep down. The impact with which he struck the sea forced water between his body and the life jacket, which then started to come loose. It also compressed his spine. Concerned that he might surface into an inferno and become trapped on his back by the jacket's buoyancy, Swales wriggled out of it so that the jacket was on one arm, while he could use the other to swim. On the surface, he was relieved to see Rae's head bobbing about fifty yards away.

The two men waved to each other and then began to swim through the swell.

•

Meanwhile the evacuation at the north-west corner was difficult, confused and dangerous, but it did progress and by 10.18 p.m. a number of men had escaped from the boat bumper into a zodiac. They included Alex Clark, assisted by 'Fergie' Ferguson, who made sure he was able to get on the rope, then went first and acted as a brake to slow and ease Clark's descent. Although he had struggled to tie his life jacket due to his lacerated thumb, Geoff Bollands also made it into the second boat-load to leave.

Back on the 84ft-level, Robert Vernon and Andy Carroll returned to the north-west corner having failed to reach C Module and switch the water pumps to automatic. The small group still gathered there decided to head down to the 68ft-level and the escape rope, but the staircase was choked with thick black smoke. Sandy Bremner, who was wearing breathing apparatus, tested the route and insisted it could be done on one deep breath; so a 'buddy' system was formed, those with BA kits accompanying those without them down the steel-grilled steps to the 68ft-level and a fresh breath. Those without BA kits were also first down the rope.

As Erland Grieve and Roy Carey both had oxygen they waited until last, having moved down to the small navigation platform. They took off their masks and tanks while Carey also removed his life jacket, intending to carry it on his shoulder. Then the explosion struck.

'Just out of the corner of my eye I saw this flame shooting along the side of the platform and I just covered up as best I could, hoping it would just be a tail of flame that would calm down and die away again,' said Grieve, who 'was enveloped in the flame and starting to scream as my hair went on fire'.

Grieve looked through the fireball to see Carey dive over the railing and, without a moment's hesitation, followed him down. In a remarkable feat, Carey had managed to go into almost an Olympic dive from 68 feet. (It was a 'performance' that did not go unnoticed. On the deck of the *Silver Pit*, drifting a few hundred yards away, Derek Ellington looked up from pulling men on board to see Carey's dive and, even among the chaos of the explosion, he couldn't help but note its grace.) He described the scene:

> It was a wall of flame more than anything else that was rushing towards us. I managed to pivot on one foot and bring my back to the flame and crouch down as low as I could. I had hoped that the flame would pass, but then I was enveloped in the flame and I realized that I was just being burned up. So I grabbed the bottom rail which was in front of me and pulled myself through and launched myself off. I hoped that I would not hit anything on my way down.

Conscious of the height from which he dropped, he tried to minimize his resistance to the water. 'I tried to make as clean a dive as I could. So I kept my hands very straight and as a result I went very, very deep.'

So deep did Carey sink that he feared he would never reach the surface before the pressure on his lungs and the diminished carbon dioxide in his bloodstream forced him to take a breath, even if it was to be seawater. When he did surface, it was to an inferno that appeared to hover eight feet above the surface of the sea where the furthest flames reached, cooking all the air underneath. It was, he thought, like being under a grill. He struggled and failed to kick off his boots and tried to keep his head underwater as his face began to burn, but it was his heart that was giving out. He thought he was going to die: 'It was a low time ... I was being cooked alive.' The waves of radiant heat emitting from the rolling flames above were igniting his hair and burning each successive layer of skin on his face, and every time he tried to cry out the sensation was like swallowing fire. After a couple of minutes Carey was unable to bear the heat any longer and, as he was convinced that his death was imminent, he decided it would be better to drown than to roast. So he deliberately sank under the waves where he fought against the shortness of breath that had previously forced him back up to the surface. He was now about six feet down and looking up through the waters at the colours of red, gold and bright orange that danced across the surface. Then a stillness flowed through his body and suddenly the only colour he could see was the white of his eldest daughter's wedding dress. Memories of the day when he walked her down the aisle blocked out the flames and he recalled the promise he made to his youngest daughter that when she too came of age, he would provide her with the wedding of her dreams and escort her to the altar. Images of her face helped to shake off Carey's 'death wish' and, once again, he began to swim to the surface and started kicking away from the platform.

It took a direct encounter with death to further rekindle Carey's fight for life. After swimming for a few minutes he came across a body floating face-down but buoyant and dressed in an orange life jacket. Carey floated beside what less than thirty minutes earlier had been an unknown colleague, and debated whether to turn him over and strip off his life jacket, knowing that while this might help to save his own life it could sink the body beyond the reach of recovery and the arms of his family. He turned and swam on.

After what seemed like an age, he was rewarded by a glimpse through the gloom of a wrecked lifeboat which, after more weary strokes on his part, was revealed to bear two men, Bob McGregor and the badly burned frame of Eric Brianchon.

Although McGregor tried to drag him aboard Carey was too weak to assist himself and instead he pulled open a part of the boat that was already cracked but above the waterline and wedged in his arm as deeply as possible, so that if he lost consciousness through exhaustion and pain, at least his head would remain above the water. Resting his head against the side of the boat, spitting salt water out of his mouth, Carey closed his eyes and began to pray.

•

In the galley, where the smoke had thickened and many men now lay on the floor, clutching to their faces torn strips of towel which had been dunked in the fish tanks, Colin Seaton struggled to exert authority and confidence in the face of growing fear. He was the man everyone was turning to for orders and guidance, but he was crumbling under the pressure.

At first he tried to tell everyone to calm down, that a Mayday message had been sent, that the whole world knew they were in trouble. Yet as the smoke became denser and the clamour for information grew, Seaton climbed up on to a table in an attempt to exert his faltering command, but his words were drowned out by the din. Mark Reid, a young man with a thin face, close-cropped hair and an earring, screamed at him that he was in charge and should get them out. All Seaton could do was to tell him to calm down, that there were four men outside with breathing apparatus trying to find an exit. Using a hand-held radio, he then tried to reach the men for an update. Four times he tried, only to be greeted with silence. Although he had announced thirty minutes earlier that they were abandoning the platform, at no time did he urge the men to attempt to make their own way out. He was waiting for a rescue that would not come.

As one witness recalled: 'The flames were right along the north face and the east face and they were actually breaking the windows and the flame just started to come in the windows and the door in the galley was on fire.'

At 22.33 hours an unknown voice relayed on channel 9 of the VHF radio what would be the final transmission from Piper Alpha: 'People majority in galley area. *Tharos* come. Gangway/hoses. Getting bad.'

7. 'HANDS TO EMERGENCY STATIONS'

'The most expensive fire engine in the world' – as Prince Philip, the Duke of Edinburgh, described MSV *Tharos* upon her launch in 1980 – was coming from a distance of 550 metres away, but painfully slowly. She was doing under 2½ knots, less than half her maximum transit speed of 7 knots, on account of the depth to which she was ballasted and the time taken to pay out cable on the two anchors that gripped the seabed.

On her launch, the editor of the *Oil and Gas Journal* wrote: 'The people of Claymore and Piper fields can certainly now sleep a little bit more safely at night.' Yet by morning, the *Tharos* would be dismissed by one survivor as 'the most expensive white elephant in the North Sea'.

The extendable gangway on which the galley's choking men had invested such hope was supported by a fire boom, a sliding structure that now creaked out at a rate of two feet every five minutes, which meant it would take 75 minutes to extend to its minimum usable length of 30 metres, and even then would require a landing spot clear of smoke and flame.

MSVs such as *Tharos* and the *Stadive*, under the control of Royal Dutch Shell and BP, were designed to be 'on scene' within 36 hours of an emergency arising, and then to provide a well-killing operation. Its hype would tonight ring hollow, while its presence would still save lives.

The vessel's principal day-to-day employment was as a dive

vessel, moving around the oilfields of the North Sea tackling whatever underwater maintenance and repairs were required. The *Tharos* was equipped with saturation diving systems that allowed divers to work in reasonable comfort for up to one month, descending each day to depths of up to 1,000 feet. Both diving and firefighting required that the structure remain absolutely stable regardless of the weather, which the *Tharos* achieved with its Dynamic Positioning System (DPS): nine propulsion thrusters, six on the port pontoon, three on the starboard, each with 2,400 b.h.p (brake horsepower), and all linked to a computer that operates them in tiny bursts in such a way as to counteract the effects of the sea and so hold the platform in position even in a Force 10 gale. Tonight the *Tharos* had been assisting with work on the new Chanter riser.

At first the men on board had felt the platform vibrate and thought that the *Tharos* must have been bumped by a boat, or maybe a heavy object had landed on the main deck. Charles Miller, who piloted the mobile diving unit, set off to sound the alarm. He was heading to the helicopter reception area on the helideck when the alarm bell began to ring, activated by one of the divers at the aft end of the vessel.

In the forward control room Anthony Ashby, the deputy OIM, told David Blair, the watchkeeping officer, to start moving towards Piper Alpha, while the ship's chief engineer said he would go and start the fire pumps. Ashby then paused the emergency fire bells to make a brief tannoy announcement: 'Hands to emergency stations. Fire on Piper Alpha.'

He then headed to the master's cabin where Alastair Letty, the OIM, had retired just half an hour before. Letty had been awakened by the alarm bells and as his cabin was at the back of the vessel on the upper deck, when he opened the curtains it afforded him a disturbing view of fire emerging from what he knew was C Module.

By the time Ashby arrived at his cabin, Letty had already decided to switch authority from the forward control room (currently in operation but facing away from the rig) to the aft control room, where he now headed. Ashby agreed to return to the front control room, brief those there on the change and transfer command to Letty, which according to the vessel's log took place at 10.10 p.m. Letty could see through the control-room windows that the blaze appeared to have doubled in size, and he was there to hear Kinrade announce at 10.08 p.m., 'abandoning radio room on fire'.

Determined to have every option available, Letty instructed that additional engines on board be started up to power the extra equipment used. He then ordered the deployment of the ship's gangway, which, unfortunately, had been retracted to allow work on the Chanter riser; mobilization of the firefighting equipment began, and he requested that all Fast Rescue Crafts (FRCs) in the area be deployed; rope baskets were lowered for expected survivors and lifebuoys and lifebelts hurled overboard. The legs of the *Tharos* were equipped with rungs, while the two forward legs had spiral staircases for descent to and ascent from the water. The hospital facility, capable of looking after 90 people and equipped with mini-operating theatre and patient-monitoring facilities such as electrocardiograph and defibrillator units, was prepared. Among Letty's first orders, however, was to get the *Tharos*'s helicopter, a Sikorsky S-76, into the air.

The helideck offered an unimpeded view but Ivor Griffith had little time to look. The duty pilot's first glance towards the rig led him to think the blaze was 'pretty controllable'. He then went downstairs to collect his headset and immersion suit, and when he returned he concentrated on preparing the aircraft for take-off, removing the engine blanks and ensuring the rotors were untied. The Sikorsky S-76 lifted off at 10.11 p.m. and orbited to the left, rising into a sky bruised dark blue and now preternaturally lit by orange and gold. By staying downwind of the smoke trails blowing off to the east, Griffith – an eleven-year veteran of the oilfields – manoeuvred the helicopter into an approach towards the blazing rig, but knew instantly that it would be one he could not complete. He then circled Piper but was forced to report at 10.13 p.m., just two minutes after take-off, that the helideck was obscured by smoke. As the Sikorsky lacked the wire winch of a search-and-rescue helicopter, there was nothing the two men could do but bear witness; so, while they continued to circle the rig at 1,000 feet, the co-pilot leaned out of the cockpit window and began to take photographs for the fatal accident inquiry they both knew would follow.

David Olley, the medic, a twelve-year veteran of the oilfields of Libya, Norway and the Arabian Gulf, rounded up volunteers, stewards from the catering company, and together they prepared the hospital, a six-bed unit complete with operating table. Attached to the hospital by interconnecting doors were four cabins, each with four berths. Beside each bed Olley set up I.V. infusions of Normal Saline and Haemaccel, oxygen as well as preparing injections of Pethidine for the pain. His plan was simple. Each person's injuries would be assessed, the most serious being directed to the hospital while the 'walking wounded' would be cared for in the vessel's cinema, manned by two experienced first-aiders. Major injuries would receive immediate resuscitation in the main hospital, then, once stabilized, would be moved out to the berths. One of the divers, an Australian who was an experienced paramedic, was despatched to set up a casualty-receiving area in the helihangar.

As Olley had not been outside since the alarm sounded, now that his volunteers were in position and their plans in place, he decided to go up onto the helideck where he expected to see a few wisps of smoke. The image of the platform surrounded by smoke and flame made him think: 'I may as well go back and close the hospital.' He did not see how anyone could survive in its midst. The seriousness of the situation was brought home to him when he overheard the helicopter landing officer, a meticulous professional, swear over the VHF radio when describing the situation to a neighbouring platform: '"'A' module is on fire, 'B' module is on fire, 'C' module is on fire. We have fire from water level to the helideck, in fact the whole fucking thing is on fire."'

•

The fire monitors were painted red, with heavy silver collars near their muzzles, and resembled cannons, appropriate as the force of the water if focused in a direct spray could punch through concrete. There were sixteen monitors of different sizes placed strategically around the vessel, seven of which (including one attached to the end of the access tower) were now being geared up for immediate use. While each was capable of throwing 40,000 gallons of water per minute over a horizontal distance of 240 feet, it was decided that to prevent the men on Piper being blasted off by the water pressure, the modules be set in a cascade mode so that the water would arc up and then flow down onto the burning rig.

The deputy OIM, James Kondol, had taken charge of organizing the fire monitors, as well as choosing the men to man the *Tharos*'s FRC and preparing the vessel's own heat shields, a screen of cooling water powered by its ballast pumps. Kondol had retired to his room during the afternoon, exhausted after working thirty hours straight, but was now alert and focused. He ordered that precautions were taken in case the ship's divers had to abandon their decompression chambers, and that water be sprayed over the fuel stores on the helideck.

The pump to power the fire modules was started up at 10.17 p.m., and it usually took two minutes until the pump was primed and the monitors ready to be opened. However, when they were turned on, instead of the emergence of a powerful jet there was but a brief, weak spurt. In the fire crew's anxiety to begin dousing the flames too many monitors had been opened too quickly, with the result that there was insufficient pressure to go round. The order went out to shut down all but one monitor, which was then used to bleed off the air.

•

The *Silver Pit*, a converted trawler with its red hull and white deck freshly painted a few days earlier, sitting 400 metres north-west of the platform, had been the swiftest to react in its role as an emergency standby vessel. George Carson, the ship's second engineer, a genial fellow with a rough beard and straggly brown hair, had just made a cup of tea and was peering out of a galley porthole when a piece of flying debris blocked out the view. Dropping the cup, he sprinted to the engine room and switched on the hydraulics that operated the winch to lower the fast rescue craft. He then ran up onto the deck and began to lower the *Nautika*, an HT 24 diesel-driven water-jet boat, capable of speeds of 25–30 knots. The crew had already assembled in their orange survival suits, led by James McNeill, the *Nautika*'s coxswain; a beefy Lewisman, 45 years old and fond of his cigarettes, he had already won over the crew with his quiet wit and genial charm, even though he had only arrived on board four hours earlier.

The *Nautika* required a crew of three, the coxswain and two others, and could hold 12–15 men, but looking out at the smoke wrapping round the rig, McNeill decided on instinct to take a fourth man. As the *Silver Pit* was on close standby the *Nautika* was in a state of readiness, so McNeill and his crew managed to make it to the north-west corner in two minutes.

Looking up at the rig, McNeill could see men in boiler suits and hard hats unrolling a hosepipe, and for a few minutes the crew believed they might not be required as the fire could soon be put out. Then they spotted, on the 20ft-level, a man trembling with fright. Mahmood Khan, a chemist, had been working in the oil laboratory before escaping down to the lower levels. His anxious demeanour prompted the crew to take him straight back to the *Silver Pit* with Charles Haffey, a deckhand and *Nautika* crewman, trying to calm him down with idle chat and the promise of a cup of hot tea, as the boat skipped across the waves.

On the *Silver Pit* Carson, who was also the ship's medic, had been down to check out his supplies and prepare beds for the injured, before returning to the bridge. The captain, John Sabourn, who was tied up working the wheel, then told him to answer the radio; it was Kinrade in the radio room of Piper Alpha announcing that he was abandoning the room. Carson told Kinrade he had received the message, then listened with growing unease to the silence that followed.

After dropping Khan back on the *Silver Pit*, McNeill and his crew headed back to the north-west corner where men had begun to clamber down the rope. In order to achieve a greater stability, McNeill managed to jam the bow between two pieces of metal structure. The next batch of escapees, which included the divers Ed Punchard and Gareth Parry-Davies, clambered aboard for the short journey back to the *Silver Pit*, where they then assisted the crew in rolling the thick rope scrambling nets so that they slumped over the side. The nets became a necessary obstacle for the next arrivals as the captain manoeuvred the ship round so that it was closer to the rig. The second batch of men carried to the *Silver Pit* had bad burns on their hands which made climbing the rope excruciating, but despite their pain they assured McNeill that they would be as quick as possible so that he and his crew could go back for their friends.

•

Carving through the now pale orange waters was the FRC from the *Lowland Cavalier*. Piloted by Ian Mackay, and accompanied by two crew-mates, he was heading towards the *Nautika*. The initial explosion had shifted the *Lowland Cavalier* (which was 25 metres off the south-west corner) off position, and showered it with burnt mechanical debris. The ship's master ordered that they abort the trenching work in which they were then engaged and move back 40 metres, dragging equipment along the seabed in their haste. At the 60-metre mark, the crew lowered the FRC, which arrived at the north-west corner at a fortuitous point, just as one man on the overloaded *Nautika* fell overboard. Mackay steered over and his crew dragged the man on-board, where he was joined a few minutes later by another man hauled from the water after falling from the platform.

With the two men shivering inside the cabin, Mackay steered the craft towards the boat bumper where a small group of survivors gathered in the water. The Tartan riser burst just as Mackay approached and his view of steel pipes, grey sea and the bobbing heads of men was replaced by a searing light as the fireball burst out and rolled over the top of the boat with the flames just above their heads. In an instant the air surrounding them became superheated. While the rescued pair were protected by the cabin walls, Mackay and his men, standing in the open, felt their skin burn and their hair singe as the temperature around them climbed above 200 degrees. The decision to grab the rope lines that ran around the edge of the boat and hurl themselves into the water was almost instantaneous. The three men, bodies submerged but with their heads amidst the heat, clung to the rope as the craft, now unguided, sped round in a full circle and once more began to head underneath the platform and towards the fiercest flames. Mackay then started to drag himself back up into the heat and over the lip of the craft until he could reach the controls, while still waist-down in the water.

Clinging to the side of the craft, he managed to put the engine on full-ahead and yank round the steering wheel so that the craft began to turn away from the platform and, finally, out to sea. Looking back, the crew saw that the men they had been steering towards were nowhere to be seen.

Once clear, Mackay headed for the *Silver Pit* where he unloaded those rescued before heading back towards the rig.

On the bridge of the *Lowland Cavalier* Michael Clegg, the vessel's master, had watched through binoculars as his FRC became engulfed by the fireball. The fire and smoke prevented him from seeing it emerge and so he considered it lost. Without a rescue craft and with high sides difficult to scale, the *Lowland Cavalier* was an unsuitable search-and-rescue vessel, so Clegg decided to position the boat 800 metres north-north-west of the platform in case anyone who escaped was now drifting downwind. Despite the distance from Piper Alpha, the waters around the *Lowland Cavalier* were soon cluttered with ejected debris.

•

The first water to fall on Piper was propelled from the four fire monitors on board the *Maersk Cutter*, a supply vessel captained by Christopher Morton which had been moored one mile northeast of the platform. At first Morton could see only a light grey dust cloud emanating from the centre of the rig, but it expanded by the second. Told of the explosion by the *Lowland Cavalier*, he saw the flames for himself once they abandoned their anchor-handling duties for the *Tharos* and began to approach the platform. Once Morton had received his instructions from Letty on the *Tharos* that he should immediately go into firefighting mode, the captain found himself in a dilemma. The vessel's FRC was stowed up near the funnels to prevent it being damaged during the *Cutter*'s earlier duties, and to haul it out and then launch it would divert his crew from the fire monitors. Yet Letty had ordered that all available FRCs be deployed. Using his own judgement, Morton concluded that fighting the fire took priority and so decided not to launch and concentrate instead on getting close enough to unleash the 10,000 tons of water per hour that would pour from the four nozzles. Within fifteen minutes the *Cutter* was 150–160 feet out from the east side of the platform and was spraying the area just above the production level. When the Tartan riser burst, Morton and his crew, watching through the bridge's full glass windows, ducked to the floor as the windows turned bright orange. They rose and with clumsy, nervous hands pulled back momentarily, before steeling themselves for further trouble and returning to their position. Everyone on-board was breathing heavily and trying desperately to process what they could see.

•

On the *Silver Pit* survivors now gathered in the large open space just forward of the bridge and designated as a rescue area. Ed Punchard was delighted to see Dick Common, a small, slight man who acted as the diver's clerk, arrive on board. Punchard asked him if he was all right and when Dick insisted he was fine, he pointed out that he had fallen onto 'the bloody boat-bumper'.

Punchard asked 'Where did you hit it?'

'Right in the middle of my back.'

Common then pulled up his shirt, but there wasn't even a mark, at least not yet. Punchard told him he had a lot of guts, climbing down the rope.

'I didn't want to go,' he explained, 'but Barry Barber said: "Don't tell me it's fucking difficult. Get down that bloody rope."'

Looking out at the rig surrounded in flames, then back at the handful of men on the deck, Punchard felt his stomach turn as the adrenalin began to drop, replaced instead by the chilling

115

realization of the plight those on-board now faced. His eye was repeatedly drawn to the accommodation module, or at least what he could see through the smoke. What on earth is happening to them, he thought. He was then distracted by Stan Mac-Leod, who said quietly: 'I'm glad you got off, Ed.'

On the bridge Walter Mitchison, the chief officer, was supervising various parts of the ship and lending a hand where he could. Most of the time he was working the VHF radio and assisting the captain. The chief engineer was in the engine room, the chef was preparing hot soup, while George Carson was in the 'hospital', a small room below deck that consisted of a couple of beds, some bandages and the barest of medical supplies.

Unwilling to just stand and watch, Punchard, with the other divers and a few able survivors, started to organize themselves. He began by heading up to the bridge to collect a notepad and pen, then started to take down the name of each survivor and the company for which they worked. As he worked his way through the group, a crew member appeared on deck with mugs and a pot of tea.

•

It took *Tharos* roughly 30 minutes from when the order was given at 10.05 p.m. to reach a close range of around 75 metres, a short journey delayed when the vessel's nine propulsion thrusters – six on the port pontoon and three on the starboard – cut out due to an overload on the power supply. The heat on deck, as well as in the control room, was fierce and eased only slightly when, at 10.41 p.m., the heat shield water spray was activated and sent a fountain of cooling seawater through jets and deluge nozzles over the front of the vessel. Unfortunately this rendered Letty, in the aft control room, almost blind, unable to see through the windows, which now possessed the clarity of a vehicle trapped in a drive-through car wash. He was already practically deaf on account of the perpetual roar of the fire that was making communication exceedingly difficult, even with those right by his side. Letty became concerned that the *Tharos* might collide with Piper's flare boom and so corrected their approach. He was also aware that although the gangway was still being extended it was becoming obvious that there would be nowhere for it to land. (Another problem was that the gangway was made of aluminium which, though lighter, weakened at a lower temperature than steel.) His plan had been to position the gangway so that it would land at the 83ft-level on the west face, a point now obscured in flames and smoke.

As the men on board *Tharos* prepared the on-board hospital, manned the monitors and sprayed cooling water on the stored oxygen cylinders, no one could hear the shouts and cries from three storeys below, where in the water an exhausted man circled the giant steel legs. Keith Cunningham, one of the divers who had leapt from the 20ft-level to escape the explosion triggered by the Tartan riser's collapse, had been swimming for over twenty minutes. He'd been following the *Tharos* on its approach and was now trying and failing to attract anyone's attention. However, he did manage to find the rungs leading up 60 feet to the deck and so, breathing heavily and shivering with the cold, he began to climb. Once on the deck, he was quickly draped in a heavy grey woollen blanket and shown down to the vessel's hospital. Taken to the door, Cunningham entered the hospital alone. When a volunteer asked who he was and what did he want, he was taken aback and said, almost apologetically, that he had come from Piper Alpha. Once the staff in the hospital realized that there was an actual survivor on-board, he was almost knocked down by men offering their assistance.

Bleeding the air from the water pumps delayed their activation, but from 10.35 p.m. onwards the water cannon began to shudder into life. The sound, like rumbling thunder, beginning quietly but growing louder, was lost amid the roar of escaping gas and its violent ignition, but the effect was visible and glorious. The water jerked out at first, punched back, then rose high into the night sky where it fanned out before falling initially into the water.

This was not what the men on Piper Alpha most wanted; it wasn't a rescue helicopter or a bridge to safety. But around 10.45 p.m., as the first drops from *Tharos* sprayed down and sizzled onto Piper's scalding metal, it was water and, for some, it would bring the gift of life.

8. 'BLOODY HELL! THIS IS SERIOUS'

The first indication that something was wrong reached 'the beach' at 10.02 p.m. when the duty officer at Wick radio station picked up a low-quality message from the *Lowland Cavalier*. It said, through hiss and crackle: 'This is the *Lowland Cavalier*. *Lowland Cavalier* alongside Piper Alpha platform ... alongside Piper Alpha Platform. They have an explosion on board an explosion on board at present ... No numbers of injured or personnel at present. Will update as necessary. Over.'

The radio station was located at the northern tip of Scotland, just 20 miles south of John O'Groats, and had responsibility for monitoring and controlling the international long-wave frequency 2182 kHz. This was the principal emergency radio frequency used to reach the distant oilfields and their various structures. The station, which was part of the maritime section of British Telecom International, immediately alerted the coastguard, logging each call as well as the subsequent messages that arrived from the radio room of Piper Alpha in the following six minutes. During this time the operator, who was on his own, was inundated with calls from other vessels reporting the rig's distress. Thus it was not until 10.12 p.m. and 10.17 p.m. that the actual messages of Piper's Mayday were written down and telexed to the coastguard, the Ministry of Defence's Rescue Coordination Centres and the insurance company, Lloyd's of London.

The responsibility for initiating and then coordinating a civil maritime search-and-rescue offshore fell to Her Majesty's Coastguard and, given Piper Alpha's location, the Maritime Rescue Coordination Centre (MRCC) in Aberdeen. The centre was fitted with a comprehensive telecommunications system, including a 24-hour radio watch on the international distress frequency, channel 16, which was carried on VHF and had a maximum range of 40 miles from the coastline where the majority of call-outs occurred. Incidents that took place further out were picked up by British Telecom International on 2182 kHz and the messages passed along to the MRCC. In search-and-rescue operations HM Coastguard worked in partnership with the Ministry of Defence, through its Rescue Coordination Centre (RCC) in Pitreavie which controlled the dispatch of helicopters and aircraft from the various RAF bases across the country.

And so, in a couple of minutes, the bad news rolled down the east coast of Scotland, from Wick to Aberdeen to Edinburgh.

In Aberdeen the watch officer at the MRCC telephoned John Wynn, the district staff officer, at his home in Bridge of Don, to inform him of the message. Wynn immediately said he would be returning to the office and told the officer to contact the MOD at Pitreavie in order to get helicopters in the air as quickly as possible.

The call was unnecessary. When Squadron Leader Geoffrey Roberts, the commanding officer at the MOD's Rescue Coordination Centre in Edinburgh, received the initial message from the coastguard – 'Mayday from *Lowland Cavalier*, explosion on Piper Alpha, position 5825N 0010E' – he assumed the worst about the situation. He decided to recall one of their Sea King helicopters which had recently taken off from RAF Lossiemouth and was at that moment heading towards a rescue incident in the Cairngorm mountains. Roberts wanted to get the pilot and navigator back to base to brief them on what little was known, and allow them to collect any extra equipment they deemed necessary. As soon as Rescue 137 received notification it began to bank left, looped round and headed back to base.

The message from RCC to RAF Lossiemouth was: 'It's Piper Alpha – there has been an explosion, they're abandoning and they're in the lifeboats. Aberdeen Coastguard would like you to proceed to the Piper Alpha and talk to them en route. The situation is a bit unclear at the moment. It looks as if it could possibly turn into a biggie.'

In a quiet restaurant in the northern town of Elgin, a few miles along the coast from RAF Lossiemouth, documentary film-maker Paul Berriff was about to start on his prawn cocktail. It was the first course in a late dinner to celebrate his birthday. He was joined by David Scott, head of factual programming at Scottish Television, and the man who had commissioned him to make a 13-part television series on the Royal Air Force Search and Rescue teams. Since 1 April Berriff, a wiry man with dark hair and a thick moustache, had accompanied the pilots and crew on each sortie, medivacking injured fishermen from treacherous seas, rescuing climbers from the Cairngorms, even whisking a sailor off a submarine at sea and taking him to Raigmore hospital in Aberdeen for treatment. When Berriff had made the initial pitch for the series, he had drawn up a list of possible scenarios for each programme. He had even pencilled in a fire on an oil rig, for programme seven.

When the pager Berriff carried with him began to bleep, he left the table and used the restaurant's phone to call the base. He was told that a helicopter was being prepped to head out to an oil platform – did he want to go? He said yes, returning to the table to tell his guests, who also included Gus MacDonald, head of Scottish Television. Berriff's sound recordist seemed initially reluctant to leave the dinner table, a glass of good wine and a rare night away from the portacabin in which the team worked when not in flight. But this was what the job was about and they couldn't let it pass. A few minutes later they were both speeding back to base, in silence.

On their arrival the 'scramble' bell was already ringing. Berriff went to grab his camera bag only to discover it had been left on board the other Sea King which had not yet returned from the Cairngorms. Instead he would have to rely on a smaller camera with a fixed 10 millimetre lens and ten tins of short end film, each containing 100 feet (roughly 2½ minutes). He dressed quickly in his dark green flight suit, picked up his helmet and prepared to board.

At home in Yorkshire, Berriff balanced his work as a documentary film-maker with his part-time role as one of the commanders of the busiest rescue boat in Britain, situated on the River Humber. He was also a trained winchman, a skill that had helped persuade the MOD to allow him access on the grounds that in an emergency he could be an extra pair of hands. The winchman tonight was Robert Pountney, an air load master, who had been on second standby and just about to climb into a bath when he received the call to come in, after giving reassurance that he had not taken any alcohol during the evening, which would have prevented him taking part.

The Sea King helicopter was the workhorse of Britain's search-and-rescue (SAR) operation. Its genesis began with a helicopter developed by the United States Navy in the 1950s for use in anti-submarine warfare. In 1975 the RAF purchased 15 airframes, with the sonar equipment for submarine detection removed to allow accommodation for six stretchers. The helicopter's powerful MEL ARI 5955 radar was retained and augmented by a Racal Decca 9447F Tactical Air Navigation System linked to a Decca Type 71 Doppler. The enhanced radar capabilities allowed the aircrew to navigate accurately in even the foulest weather conditions, permitting rescue missions in restricted visibility. The Racal Doppler system also allowed the helicopter to fly without updating its navigation systems from external sources, which would be out of range on long-distance SARs. Each helicopter had a pilot and co-pilot who were assisted by the Louis Newmark Automatic Flight Control System (AFCS), which provided a three-axis stabilized flight (pitch, yaw and roll) as well as allowing an easy transition into a hover. The crew was completed by one radar/wing operator and one winchman. The cabin could hold 17 seated survivors or six stretchers. The Sea King had a maximum speed of 250 km per hour and a radius of action of 510 km.

At 10.22 the first Sea King helicoptor, call sign R137, bright yellow with a black and white duck painted on the main door, lifted off from RAF Lossiemouth. Six minutes later a second Sea King, call sign R131, took off from RAF Boulmer in the north of England. It had quickly become apparent that communication at such a distance would be exceedingly problematic since the helicopters would be outside normal air traffic communications. So at 10.19 p.m., the RCC instructed RAF Kinloss (20 miles further west along the coast from RAF Lossiemouth) to scramble a Nimrod aircraft. The Nimrod, given the call sign Rescue 01, was to act as a flying communications platform, handling the signals from each helicopter and the various vessels, reporting back to the RCC using HF transmissions. It had the fuel capacity to circle the platform for eight hours.

When the maritime rescue coordination centre in Aberdeen was alerted by the rescue control centre in Edinburgh to the presence – at 50 degrees north, 3 degrees east – of the Standing Naval Force Atlantic, a flotilla of frigates and destroyers from seven nations then in the midst of a military exercise, they immediately requested that the force be diverted and make haste towards Piper Alpha. The request was made at 10.35 p.m. The largest maritime rescue operation since the Second World War was now under way.

By then R137 had been airborne for 13 minutes and was flying at an altitude of 10,000 feet. When the exact messages from Piper Alpha were telexed from Wick radio station to the coastguard and rescue centre, they were then passed on to RAF Kinloss where a mix-up occurred. In the early minutes of the flight the crew were informed that the men on Piper Alpha had been 'evacuated'. On board, Berriff and the crew began to relax slightly; an evacuation meant a controlled act, properly supervised. However, a minute later the error was corrected; the rig was being 'abandoned'. Berriff turned to the sound recordist, saying: 'Bloody hell. This is serious.'

At the maritime rescue coordination centre in Aberdeen, a feeling of frustration had begun to grow. The control room consisted of a long row of desks, on which sat communication equipment and heavy grey plastic phones, staffed by officers in starched white shirts and epaulettes to designate their rank. The air was often heavy with pipe smoke. Yet too often when the phones rang (and they didn't seem to stop), it did not result in hard facts on the current situation but instead an awkward conversation with a newspaper or television reporter who seemed to know as much as they did, which was very little. Shortly after Wynn arrived, the first call he answered was from a French press agency. He then ordered that a separate line be set up for press calls only.

When the MRCC had run offshore emergency drills they featured a smaller fire, or explosion, which then escalated while still leaving the OIM access to communications. A scenario in which they had no communication with the rig had not been run. The *Offshore Emergency Handbook*, prepared by the Department of Energy and with which all the oil companies operating in the North Sea were familiar, specified that in the event of the OIM abandoning the rig, the role of On-Scene Commander (OSC) should fall to the most suitable candidate capable of carrying out the role. According to procedure the MRCC was supposed to liaise with the oil company to designate the OSC, who had a number of crucial tasks.

According to the handbook the OSC was responsible for

> executing the plans of the search and rescue mission co-ordinator . . . modifying those plans as required to cope with changing on scene conditions; assuming operation coordination of all units assigned by the coordinator; establishing and maintaining communications with the coordinator; submitting situation reports at regular intervals to the coordinator for action; establishing and maintaining communication with all facilities; performing search, rescue or similar operations; providing initial briefing . . . coordinating and diverting surface facilities or helicopters and aircraft to evaluate sightings; and obtaining the results of the search as each facility departs the scene.

As there would be men in the water it was also necessary to appoint a Coordinator Surface Search (CSS).

The problems with speaking to the *Tharos* or any of the vessels positioned around the rig flowed in two directions where they wedged in the bottleneck of Wick radio station. First, the station was bombarded in the first hour or so with messages from dozens of rigs, platforms, support vessels, fishing vessels and even aircraft, offering assistance and further information, with each call requiring to be logged and any extra information extracted. Second, these calls inhibited the coastguard relaying through Wick to speak to the captain of the *Tharos*.

To do his job, John Wynn needed to know the nature of the explosion, the number of persons on-board, what the OIM was doing, what the weather was like, what other vessels were in the area (and what they were doing), what communication facilities were available, as well as life-saving equipment. He needed to know if this was an 'alert' situation, or a 'distress' situation where there is grave and imminent danger. With dozens of questions but no answers, he had to assume the worst – that everyone had abandoned the platform by any necessary means. It was, as he would later describe, 'an hour of chaos'.

Wynn and his staff at the MRCC concentrated on seeking information from Occidental's headquarters in Aberdeen, but the office was not manned 24 hours a day. The phone was answered by a security guard, who was in the midst of calling in senior staff and ignorant of what was going on beyond the banal description of 'an incident'. (There would later be a dispute over whether the MRCC did persistently try but fail to be put through to the *Tharos* via Wick radio. The radio station staff insisted they had no record in any of their log of calls from the MRCC requesting a link to *Tharos*. It was suggested that when the MRCC eventually did make contact with the *Tharos* it was not through Wick radio but Stonehaven radio, who dealt with day-to-day communications with the *Tharos*.)

•

At 10.45 p.m. a third helicopter, R117, took off from Shetland while a fourth rescue helicopter, the second Sea King, R138, took off from Lossiemouth at 10.51 p.m. The Nimrod aircraft, Rescue 01, recorded wheels up from RAF Kinross at 10.55 p.m. It had not been airborne long, had climbed to 20,000 feet and was still almost 80 miles from the scene when the pilots saw a faint orange glow on the horizon.

Rescue 01: 'We're expected on the scene in two zero minutes and we have the rig visual at this time.'

RCC: 'Rescue 01, this is Edinburgh rescue, situation confused at this end. Request frequent sitreps [Situation Reports]. Over.'

Rescue 01: 'Roger. We will pass frequent sitreps, as yet we are still in transit to the area and we can actually see the rig on fire and stand by for further transmissions.'

9. FEEDING THE FLAMES

The night was so clear that the men on the Tartan platform, 12 miles south-west, could see 'a red envelope of flame' projecting from the north side of Piper Alpha, just below the production modules. A few minutes after the platform's OIM, J. Leeming, heard the Mayday message he told the Production Supervisor, Michael Moreton, to monitor pressure on the gas pipeline that ran from Tartan to Piper and then on via MCP-01 to St Fergus. Leeming then called his own supervisor at Texaco's headquarters in Aberdeen.

Between 10.10 p.m. and 10.20 p.m. Moreton discovered that the telemetry system had frozen as from 10 p.m., which resulted in the VDU display updating information only from Tartan, not from Piper. In order to discover what was going on he tried to call both Piper and Claymore on the omnibus system with no response, yet it was decided to maintain production in the belief that Piper was doing the same.

He did notice, however, that the pressure in Tartan's gas pipeline to Piper was now climbing, which indicated to him that the import valve on Piper had shut, leaving the gas to build up in the pipe. When Leeming was informed, he instructed Moreton to shut down the export compressors – in effect turning off the tap – and then close Emergency Control Valve 54, which was in accordance with the procedure in the event of a serious emergency on Piper Alpha. Moreton and Leeming would later insist that the decision to shut down the export compressors was made at 10.15 p.m., but the actual closure of ECV 54 was not recorded until 10.25 p.m.

Five minutes earlier, at 10.20 p.m., the Tartan riser on Piper Alpha exploded. When Moreton was told he rushed to a window and saw the fireball envelop the rig. Back at the controls, he also noticed that there had been a sharp drop in the pressure inside the gas pipeline between 10.20 p.m. and 10.25 p.m. He thought this odd, and discussed it with his colleagues, but did not make the connection with the recent explosion.

The error could be explained by the fact that the pressure chart for the gas pipeline to Piper was still showing a horizontal line, indicating a consistent pressure. They did not know that the sensor in the pipe which provided information for the computer chart was upstream of ECV 54. There was a second pressure gauge downstream of ECV 54, but this was not usually monitored.

On the Claymore platform, 22 miles west of Piper Alpha, S.B. Sandlin (the OIM) could not see the flames. The Mayday message from their sister rig was passed on to him immediately after broadcast and although he believed a major emergency to be in progress, it would be almost an hour before he shut down all production, despite four separate requests from his deputy. The cost of shutting down unnecessarily would be millions of pounds in revenue and a black line through your career. Sandlin had been on-board the Claymore back in 1984, the last time when Piper Alpha had an emergency, and there had been no need to shut down then. He decided to wait.

With his operations superintendent James Davidson, Sandlin then tried to telephone Piper Alpha but discovered the line was down. When the second Mayday was broadcast, he first ordered the Claymore's own standby vessel, the *Nautica*, to head directly to the rig. He then tried to telephone the production and pipeline superintendent, J. Bryce, back on the 'beach' at Occidental's headquarters in Aberdeen, in order to report what he knew; but that line was clogged with traffic. At around 10.15 p.m. Davidson managed to get through to the helideck officer on the *Tharos*, using the VHF. He was told that Piper was on fire at its production modules on the west side, and that a large volume of black smoke was blowing over the helideck from the east side.

Davidson told Sandlin what he had learned, and said that they should shut down the main oil line. He was worried about the risk of oil being released on Piper as a result of heat failure of the pipes leading to a blowback. He also wanted to start depressurizing the gas pipeline between Piper and Claymore and begin flaring off the gas. The longer the fire went on, the greater the risk of a pipe failure. Sandlin was already aware that Piper had stopped oil production, but having checked on the monitors and found that the pressures in the pipeline were stable, he decided that Claymore's production should continue as normal for the time being.

Sandlin ordered that the pressures in the pipelines be closely monitored with any changes reported immediately. Yet there was a problem, in that the telemetry systems providing information from other installations had failed. The operators now had to note what was shown on the pressure gauges for the gas pipeline and look at the chart recorder in respect of the oil pipeline.

At 10.20 p.m. the telephone system failed while Sandlin was attempting, once again, to contact the Emergency Control Centre in Aberdeen. Once more Davidson tried to argue for production shutdown, but Sandlin disagreed. In the space of the next ten minutes the *Tharos* broadcast an update describing how the fire was spreading and reporting that people were in the water. Davidson then headed up to the helideck from where he could now see a glow on the horizon in the direction of Piper Alpha.

At 10.30 p.m., Davidson tried a third time, but Sandlin argued that he did not think the situation on Piper was beyond the control of its fire pumps. He did not know these were inoperable.

Sandlin's decision to continue production was based on the maintenance of pipeline pressure and on, what he considered to be, a limited knowledge of the situation on Piper Alpha. Although aware that the situation was worsening, he later said it was 'still not indicating to me that a major disaster was in the making'. He was relying on his own judgement and the contents of Occidental's *Pipeline Operating and Emergency Procedures Manual*. A key paragraph stated 'If it is immediately clear that a major problem exists such as the rupture of the pipeline or a serious incident at the platform, shutdown of the platform or the whole system will be initiated by the affected platform. Each location can only initiate shutdown of its own systems so it is vital to inform the other locations of the situation and of the need for action so that they can initiate their own shutdown actions.'

He based his judgement on the impression that Piper Alpha had maintained production and that the Claymore pipeline was secure with the pressure reducing gradually through normal usage. Had the OIM on Piper Alpha contacted him, he would have immediately shut down production.

10. THE LONG DROP

Mark Reid did not want to die surrounded by strangers, but as the smoke had thickened in the galley, where the only light came from the disturbed dance of a few torches and the flames outside the windows, this is what he believed would be his fate.

There were about 100 men dressed in a mixture of orange survival suits and life jackets, jeans and sweaters, boiler suits and boots – standing in groups, sitting on chairs or lying on the floor where the air was clearest, all waiting to be rescued and displaying a mixture of patience and panic. Those who began to scream and shout and demand that something, *anything*, be done, were calmed by those who bit down on their fear and were able to offer the worthless assurance that help would soon be at hand.

The rising temperature in the room, plus the smoke burning his throat and blinding his eyes, had brought the panic in Reid to the boil. He tried to cool himself down by stumbling over to the drinks machine, barely visible in the gloom, where he got a polystyrene cup of blackcurrant juice and poured it over his face. Unfortunately it did not assuage the fear that had begun to grip him; he was now hot, scared and sticky and quickly lost control, running around the room and banging into people 'like a chicken with its head chopped off'. But in between his tears, rants and calls for help, Reid realized the futility of his actions and eventually fell into a weary despair. If he was to die, he reasoned, he did not wish to do so alone; he wanted to be holding the hand of a friend.

He then began to shout: 'Is there anyone here from Bawden drilling?'

There was, but they did not hear. Fred Busby, a driller with the same contractor, was sitting on the floor of the galley, wheezing as the smoke began to take effect. Although close to the ground, he was still breathing it in and now felt giddy. He knew that if he stayed any longer he would collapse.

At this point John Gutteridge was picking his way across the room, stepping in between the men, until he found his way to where Busby was sitting. Crouching, he said: 'If we don't get out of here soon we are all going to die because of the effects of the smoke.'

Busby replied: 'Let's see if we can get back to the rig floor, John.'

•

Smoke, hot gases and flame were spreading into the reception area on D deck through a doorway from the Living Quarters West (LQW), as well as a second door on the south side of the building which gave access up to the helideck. These conditions could not be survived for long, but closed doors leading off the reception area protected the rooms beyond, at least for now. But the smoke was also spreading to those rooms near the reception area, as well as the kitchen storeroom by way of the voids in the ceiling. The platform's loss of power was fortuitous in one respect – it had cut off the ventilation fans which, had they been active, would have sucked in further smoke. Yet hot dense smoke was also coming through a fire-resistant door between the reception area and the staircase. Because this had been hooked open to ease the movement of personnel, smoke was now spreading into the passageways between the reception area, the dining room and stairwell. Since the door between the dining room and the passageway was repeatedly being opened and closed the smoke penetrated further; it was also entering the dining room from the kitchen, which in turn was being fed by the ceiling voids above the storeroom. The worst area was the north corridor of C deck, which was being flooded with smoke and hot gases from the LQW. The east and north faces of the ERQ were now under severe attack from the fire, which had also spread to the end cabins of the lower three decks.

The picture in the Additional Accommodation West was much worse. The building's faces and roof bore the full ferocity of the fire with the temperature rising above 1,000 degrees, pushing through the walls and igniting the rooms, while fire and hot gases continued to sweep through the external doors and fed into the ventilation system. Without water passing through to keep it cool, the copper piping of the sprinkler system was melting.

The consequence of the fire was to create a hostile environment, ripe for confusion. To the unfamiliar, the interior of Piper Alpha with its internal stairways and uniform corridors was not an easy place to navigate; in darkness, with the air shaded grey by smoke and the walls pulsing with heat, it became at best extremely difficult for some and at worst impossible for many others. For the first forty minutes after the initial explosion men were struggling along the corridors and up and down the stairwells, which as time progressed became increasingly congested with those who could only sit and wait.

The corridors were just a few metres wide and seemed shrunk with the claustrophobic effects of the smoke whose grey, choking filter turned the familiar picture into stark screens of static. Small groups of men held on to each other's shoulders with one hand while using the other to hold up soaked towels and cloth strips over their mouth and nose. Heads tucked down onto their chins, they coughed and choked as they moved, slowly feeling the walls in search of an exit and scalding their hands on the metal handles of those doors that held back a fire. The conditions were akin to a blizzard, one where the temperature never dropped but instead inexorably rose, and in their rubber survival suits each man began to sweat profusely and thirst for water. While those in groups had company to keep the panic at bay – at least for a few more minutes – those who found themselves alone in the smoke were quickly stripped of their emotional defences. A man becomes disoriented quickly in dark smoky places. Each time one of them turned round and still did not know where he was the screws of fear and despair turned tighter. In such a situation, any man thinking of his wife and children and what they are doing at that precise moment can fuel the rising panic as much as strengthen his resolve. In those corridors, prayers were uttered and curses were hurled, but nothing altered the fire's relentless progress and the countdown of the clock.

In the galley the cod, skate and dogfish were content to swim around the elongated aquarium despite being disturbed by men who continued to soak their towels in the tank. Mark Reid had still been calling for company when his cries were answered by a torch beam of light and a voice calling him over. It was Jeff Jones, a fellow driller, who was prompt in telling him what Gutteridge had said to Busby: that they had to get out or die. Together the pair pushed their way through the crowds to the back of the galley and into the kitchen, where they were faced by a double door. A small group of men who had gathered there warned that they would be badly burned if they pushed on through and out of the door, but Reid shouted that there were already people dying in the galley, so what was there to lose? When he and Jones pushed the door open and stepped outside, they were surprised to find a few men already there, sheltering in a steel container used as a cold-storage unit for kitchen supplies. As both of them also stepped inside to catch their breath and escape the heat that had hit them, the chef's supplies were being ransacked and Reid was handed three or four tomatoes. He squashed them over his face and neck before sucking the remnants in a bid to alleviate his parched throat.

Meanwhile, a little group had gathered with Busby and Gutteridge. Together they succeeded in pushing their way out of the galley and over to the internal staircase where the smoke had shrunk visibility to the point where they had to shout to each other to keep in close contact. In the smoke and dark, time seemed to stretch until it was as taut as their nerves. Eventually the minutes and metres passed and the group made their way back down to the Bawden office, which had a door that opened on to a lower deck where there was a barrier of smoke in which red embers glowed. When the men edged out of this door, they reached an escape ladder that rose up the side of the accommodation block and towards the tea-shack, but a volunteer who climbed as high as he could returned and reported that the fire above was just as much an obstacle as the smoke below. They had noticed that the conditions of the fire could change rapidly. Fierce flames could briefly quieten down; while an area which appeared clear, but was in fact swirling with invisible gases quietly boiling towards their ignition point, might suddenly flare up. As the group gathered in the doorway, it was Fred Busby who steeled their nerves.

'I'm not going to stay here,' he said, then added: 'Let's go outside.'

When he moved out he found that he could see between 5 and 10 feet, and noticed that there were two more people standing outside the doorway.

Turning to Gutteridge, Busby said: 'John, we'll make our way to the drill floor.'

'We don't know where we're going,' Gutteridge replied.

To which Busby responded: 'Well, we'll just keep our heads down and give it a go anyway.'

Back in the galley Jim McDonald – the rigger who had helped to lower PSV 504 to the ground less than six hours before – approached Alan Carter, the lead production operator, and asked what was happening. Carter didn't answer and appeared addled, which led McDonald to assume he had become delirious from the smoke. The encounter left McDonald – a large man with a bald head and mutton-chop whiskers – shaken but with the clear understanding that deliverance rested in his own hands. Thinking to himself, 'Get yourself off,' he went to collect his friend and cabin-mate Francis McPake, who was reluctant to follow. McDonald managed to cajole him as far as the reception area next to the bond shop, where the smoke drifted around the boxes of Marathon bars, packets of Fisherman's Friends and Murray Mints.

Here McPake suddenly stopped. 'We've done our muster job. They'll send choppers in,' he said.

'I've tried to speak to Alan Carter; but he cannot talk to me, Francis. There's something drastically wrong on this rig. We'll have to get off,' McDonald replied.

The words did no good; McPake would go no further and instead slumped down onto the floor, where McDonald – strug gling between loyalty to his friend and the demands of personal survival – left him with regret and remorse.

Having found his way to the stairway, he began to crawl down level after level. The smoke was so thick that he could hardly breathe, so he put his jumper over his head so as to strain the smoke through the wool. And then there were the bodies. Some men had passed out after 15–20 minutes and parts of the staircase were blanketed with their bodies. McDonald, in deep distress, was now forced to crawl over them.

On the edge of the pipe deck, Gutteridge and Busby crouched

down and set off by feeling the metal diverter, then made their way across the deck followed by the rest of the men. As they moved Gutteridge looked to the west where he could see a chemical tank with its paint ablaze and releasing billows of toxic smoke. The group also swelled in number to about eight as other men on the pipe deck spotted them and sought, if not the illusion of safety in numbers, then at least the comfort of company. The smoke of the pipe deck was taking its toll on one man, Adrian Powell, a 28-year-old crane operator, for whom each breath was 'like somebody grabbing your throat tighter and tighter'. However, the smoke eased a little once they reached the next stage of their escape route.

The drill floor was surrounded by high steel walls to protect the workers from the winds that could whip through and the tremendous heat generated by the twin flares, so Gutteridge and Busby climbed up to appraise their situation. From the top of the wall they could see that everything to the north was on fire, the accommodation modules and all around them ringed in flame, with black smoke spiralling up into the pale blue sky. Gutteridge saw flames rise above the helideck and knew there would be no airborne rescue. Busby's eye was drawn to the east-side crane, which was burning like a candle dipped in petrol. They then climbed back down and joined the other men in collecting life jackets and lengths of rope from the 'dog house', a small shack that on a better night would have offered hot coffee and scalding patter. The group then pressed on, down a short staircase to the skid deck, which offered plentiful evidence of the fire burning in the modules below. Scalding grey smoke was rising out from under the hatches and the steel floor was 'like walking across the top of a boiling kettle'. The handrails were burning to the touch. When Gutteridge looked down into A Module he saw a distinct fire to the side of a well, and reasoned it must be a gas fire on account of its clear white glow and jet of flame.

Explosions seemed to be going off everywhere, the men crouching down in alarm every few feet. They were also nervous about the supplies of oxygen, acetylene and nitrogen that they knew were stored on the west side of the skid deck, while all the time the temperature continued to rise. Busby was anxious that they should find a route into the sea on the east side; he could see which way the sea was running that night, and didn't want them getting into the water only to be carried under the platform and cooked. Yet every route to the east was blocked by a cluster of small fires and jets of flame, the only clear path being towards the south.

The exit out was via a 5-inch hose. When the group reached the south-west corner, they climbed down to a navigation beacon and from there to the 68ft-level where, just a few yards along the west side, Gutteridge spotted the hose, which had been tied off and now dangled down, coming to a halt roughly thirty feet above the sea. A rescue craft had been spotted in the waters below and suddenly hope began to bloom. The men, now numbering about a dozen, lined up in readiness to take their turn, lowering themselves arm over arm before dropping into the sea. Most were anxious to go but one man, clearly terrified at the prospect of a long drop into the open water, kept passing and handing his place on to the man behind.

Gutteridge found himself deliberately trying to take in all that was going on around him – the roaring noise, the burning heat, the orchestra of mini-explosions which rose above the persistent blow-torch thrum that seemed to make the very air vibrate. He was the fourth or fifth man on to the hose. Busby went ahead, while Powell made a point of looking down before he dropped from the hose in case he should land on anyone else.

Once in the water, each man swam over and held onto a rail by the steel leg, hoping to be spotted by a fast rescue craft.

A few hundred yards out, William Flaws, the coxswain of the *Tharos*'s FRC, was trying to avoid being distracted by the grim spectacle rising up before him and instead train his focus on the waterline. The crew was already finding it difficult to distinguish the orange life jackets from the various similarly coloured pieces of detritus propelled off the platform, and he was anxious not to miss anyone. Scanning the scene, he spotted two men climbing down what he thought to be a rope and powered the boat over, positioning it just below them. His crew dragged the pair on board and he then manoeuvred over to where Gutteridge, Powell and Busby were clinging to the platform leg. Just as the boat was moving towards one more person in the water, it seemed to lose power (caused, they would later discover, by a loose fuel connection). The man was picked up and Flaws managed to restart the engine, but he was anxious to get them all back to the *Tharos* and reluctant to continue the search, even though he still had space on board, just in case the engine should fail again and leave them all at the mercy of water and flame.

A few minutes earlier the fast rescue craft from the *Sandhaven*, crewed by Brian Batchelor, Malcolm Storey and Iain Letham, had seen the *Tharos*'s FRC move towards the men, so they decided to head on up towards the north of the platform in search of survivors. Unfortunately the smoke was so thick and clawing that they could see nothing, in between their coughs and splutters; so the coxswain, Iain Letham, spun the craft around and made towards the south-west corner where they knew there were men in need of rescue, as well as the visibility to succeed. As they returned they could see from a distance of about a hundred yards that the FRC was heading back towards the *Tharos* and its crew waved them on towards those survivors still gathering at the corner. Letham then opened the engines, the *Sandhaven*'s FRC picked up speed and barrelled towards the corner.

The time was a few minutes before 10.50 p.m.

At this moment Bob Ballantyne was looking down towards the approaching FRC from 20 feet above the waves on the walkway of the cellar deck. How he came to be here, alone, was because of an instinct that in one way was right and in another wrong, and even the cajoling of close friends could not alter his course.

Until a few minutes ago, Ballantyne's team had stuck together. He, Charlie McLaughlin and Ian Gillanders had searched the various levels of the accommodation modules together in case anyone was trapped or still asleep. As they found no one unaware of what was going on, they returned to the galley, but left when Ballantyne realized that the *Tharos*'s helicopter could land in seven minutes and if it couldn't make it, neither would any 'birds' from the beach. The group now numbered roughly a dozen, some of whom were smart enough to feel the heat coming off the steel bars that ran across the front of the doors before they tried to open them. Like others before, they exited the accommodation block through the office level that led out onto the production level, where the full extent of the fire was apparent and frightened a few of the men back in and up to the galley. Ballantyne tried to talk them out of their retreat – 'You've got to be jokin' about getting a heli' out of here' – but proved unsuccessful.

Those who remained then split up, having first agreed that they would return to this central point and alert the others if they managed to find a viable escape route. It was a pact which Ballantyne knew individuals would be unlikely to keep; in the current conditions, if you could get out you would. As they moved along the pipe deck beside the burning east-side crane, they looked up towards their designated lifeboats, now unreachable behind a blazing barricade. At one point fireballs – or so Ballantyne would swear – came firing along the gangways as if shot from a flaming cannon. The men just managed to dive into a fabrication workshop known as the 'White House'. Once they emerged, a few of the group decided to climb scaffolding which led up to the helideck and, despite his doubt on the efficacy of helicopters in the crisis, Ballantyne briefly considered joining them before dismissing it as hopeless.

It was just then that Les Morris returned with some of his party, saying that the best route would be to make their way along the gangway and through the drilling derrick. The group then took the stairway on the south side of the drill floor down towards A Module. Ballantyne was second to last coming down the stairs on to the production level, by which time the rest of the group had already decided to head west, towards where the *Tharos* was positioned. But Ballantyne balked. He could see the blowout preventers up ahead, and they were melting. Along with the terror this induced was a memory of the surrealist works of Salvador Dalí and his melting clocks. Even in the midst of the inferno, Ballantyne's passion for culture and learning could not help but put the horrors he saw within a cultural frame. When Charlie and Ian noticed he had failed to follow, they returned to cajole him. Each time he looked ahead a quiet voice said: 'No.'

Charlie said: 'C'mon, wee man.'

Ballantyne could only reply: 'No, I don't trust it.'

And so they parted, and would not meet again.

While his friends went west, Ballantyne and another man moved east and eventually reached the south-east corner at the 68ft-level. There was no stairway down to the 20ft-level cellardeck and while his companion – whose name he would never learn – was able to climb down the leg itself, Ballantyne (who was indeed a 'wee man') could not reach the first handrail. Having tried to climb down from the other side with the same lack of success, he then looked around for anything that might assist him. Finding some rope nearby, he tied it round his waist, trying to remember how the mountaineers wore it to avoid rope burn, then lashed it to the handrail and lowered himself down until he could get a firm grip on the rail. He then moved over to grip the cathodic protection cables, and used them to aid his climb down. The evening was still light and when he reached the gangway he felt relief, despite the heat from the flames roaming over the underside of the platform. Then he noticed a cluster of people, a few hundred yards away at the far end of the cellar deck. His heart lifted as he began to walk towards them: he had spotted Charlie and Ian.

In the storage container outside the kitchen door, Mark Reid had finished his tomatoes and rested for a few minutes, sitting with the other coughing men on wooden crates and giant tins of cooking oil. The container was just below the helideck and when Reid walked back out, he could see Douglas Findlay, a mechanic, with Bawden who was dressed in jeans and the company's safety sweater, and who shouted down that you could 'catch your breath up here'. He then moved back from the edge of the helideck and out of sight. Reid could also see that there were already three or four men dressed in survival suits standing on the roof of the radio room, which was a few feet above the level of the helideck. Since for the moment the air did appear to be a little clearer, Reid and Jones tried to make their way up.

As they looked for a suitable route the pair came upon a firehose. Since they had come out from the galley, one specific concern of the men outside was the fires creeping towards the helifuel tanks that were situated on their level. Reid and Jones turned on the hose, then threw it down when the water failed to flow. They climbed on to the top of a small storage unit roughly six feet below the helideck, with the idea of jumping up from this point and grabbing hold of the lip of the radio room roof before hauling themselves up. The adrenalin was pumping and anything seemed possible to achieve a clean breath of air, but even this wasn't enough to provide the extra elevation required. They retreated, with nothing to show for their efforts but burnt hands from slapping but failing to catch on to the scalding steel. Reid and Jones returned to the container. Their exertions in the intense heat had left them severely dehydrated, and Reid was once again beginning to lapse into despair.

Then it arrived. 'Rain' from the *Tharos*. What had seemed to begin with just a few drops quickly became a heavy shower that, for a few brief seconds, brought something resembling joy back on-board. Reid tried to catch it in his mouth, then collect it in his cupped hands, but neither method enabled him to gather enough to quench his thirst. And so, when he saw that the water had pooled in puddles on the floor, he got down on his hands and knees and began to lap like a dog.

As one man kneeled, another stood with arms outstretched and allowed the spray to saturate him. Ian Fowler, a 30-year-old joiner with the Wood Group, found himself revitalized by the water, baptized with hope. Once drenched, he moved away from the food container and saw through breaks in the billows of smoke that there were men on top of the Additional Accommodation West; and so, assisted by a few others, he managed to clamber up. From there he could see through the veil of water to the vast coffee-table hulk of the *Tharos* and the individual men on the decks. He waved to them, and they waved back. When he turned and looked down to the north-west corner he could clearly see a body floating face-down and naked, the clothes having been blown off in the blast.

The helideck had briefly become home to a huddled community of the trapped, those who had climbed up instead of down and were waiting for salvation to come from the sky. There were small groups dotted across the two helidecks as well as the roofs of the Accommodation West and the radio room. The only person on a higher point was the man scaling the drilling derrick in the hope of rising above the clouds and catching a helicopter out. He was spotted by the crew of the *Tharos* who transmitted a message to Ivor Griffith, their helicopter pilot, who was just returning from refuelling the Sikorsky S-76 on the Claymore. Griffith was told that there were people on the derrick: could a rescue operation be attempted? His initial doubts were unfortunately confirmed by the time he returned to find the derrick muffled in dense black smoke.

On the helideck, some men stood in small groups while others stood on their own, like single statues, rooted to the spot with fear. A few paced continuously from side to side in the hope of spotting an escape route they might have missed, but knew they had not, all the while looking up into the smoke-filled sky in the hope of seeing the white spotlight of a rescue helicopter. The deafening din of the escaping gas, roaring all around, and the sharp crackle of fire meant that they would not hear its approach. At times they were entirely ringed in smoke, then the wind would blow out a window through which they could see the *Tharos* only a few hundred metres away, but a distance that might as well have been miles for all the assistance they could offer. The space between the men on the helideck and on the *Tharos* below appeared so narrow when they exchanged waves, but then the billows of smoke would cover the 'window', sealing them off once again.

On the roof of the radio room a group of men began to talk as they had never done before. The banter and cruel japes of the ordinary working day were removed from their vocabulary. Instead they spoke of how much they loved their wives, the fear of never seeing their children grow up – and where, in their darkest hour, was God? Joe Meanen stood and listened. He had no wife, no children, but was determined to have the option and somehow survive the night. Then he walked across the roof and on to the emergency helideck with the idea of attracting the attention of the men on the *Tharos*, hoping they would lower down the crane jib so that he could grab hold and get off. Mark Reid, who was among the group, was shouting: 'Come in! Come in!' A few of the others followed Meanen almost to the lip of the deck, and he was looking down towards the *Tharos* when another gas riser, weakened by the intense heat, gave out. This flooded the fire with enough gas to power millions of Britain's homes, enveloping the helideck with sheer walls of fire that rose up 500 feet above their heads, curving over to form a basilica of flame.

In the white-hot heart of the inferno the gas riser, which ran from the Piper 22 miles along the seabed to the MCP:01, had become blackened, and the steel's strength weakened as the temperature rose above 700 degrees. The fire was like a blow torch burning through bars to free a friend. And at 10.50 p.m. the pipe gave up. From a distance the explosion was like the birth of a star as 51 million metres of standard cubic feet of gas (the equivalent of 1,280 tons – three times the quantity present in the explosion at 10.20 p.m.) erupted around the platform.

Seconds before, the *Sandhaven* FRC had become stuck. After waving off the *Tharos*'s FRC as it returned to the vessel, the crew had manoeuvred themselves into position by the leg at the southwest corner and were benefiting from the curtain of water now coming from the *Tharos*. Men were climbing down ropes as well as the fire-hose and the crew picked up four straight from the platform, then three more from the water. Just as they were pulling away, Iain Letham saw two more men climbing down from the platform and so they spun round 180 degrees, carving white foam as they went and scooped them up. There were now

The Piper Alpha platform, with its drilling derrick rising from the centre, was once the single largest oil producer in the world.

A worker on scaffolding on Piper Alpha
as the flares burn behind.

Sue Jane Taylor's photograph
of the platform's duty-free shop
and its attendant would later be
transformed into a painting.

The helideck, the principal means of access to the rig, was 175 feet above the water and from here a number of men leapt: some to their deaths, others to safety.

The 'spider' deck, 20 feet above the waves, allowed men access to the sea and rescue boats. The cut-off hollow tube on the right-hand side is a 'boat bumper'.

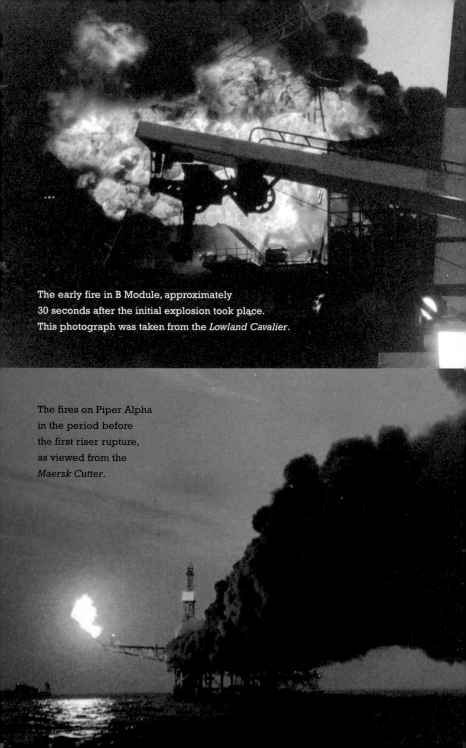

The early fire in B Module, approximately
30 seconds after the initial explosion took place.
This photograph was taken from the *Lowland Cavalier*.

The fires on Piper Alpha
in the period before
the first riser rupture,
as viewed from the
Maersk Cutter.

Fire in the night: Piper Alpha is consumed by flames after the rupture of the Tartan riser at around 10.20 p.m.

Despite the searing heat a fast rescue craft (FRC) moves towards the 20-foot level, after the first riser rupture, in an attempt to save those submerged but burning under the flames.

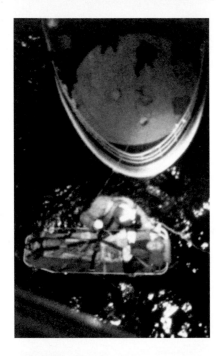

Rescue 138 winches the first of
the survivors, Eric Brianchon,
off the deck of the *Silver Pit*
as Ed Punchard looks on.

As dawn breaks Rescue 138
returns to the *Tharos* with
two victims found dead in
the sea around Piper Alpha.

The morning after: all that remained of Piper Alpha on the morning of 7 July was the buckled and blackened structure of A Module as gas lines continued to burn.

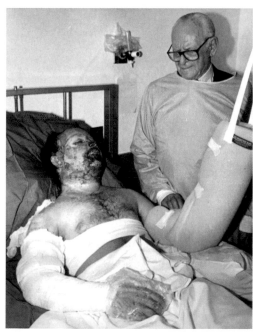

Visiting the wounded: Armand Hammer, chairman and CEO of Occidental, comes face to face with the human cost of the tragedy.

Raising the dead: the accommodation module of Piper Alpha is lifted from the seabed.

The monument to the dead of Piper Alpha, sculpted by artist Sue Jane Taylor, not only bears the names of the 167 men who perished, but also contains the ashes of unidentifiable remains.

three crew and nine survivors on the *Sandhaven* FRC, but most unfortunately there were also coils of rope whose excess length had become entangled with the engines. As Iain and the crew set about untying the ropes, the entire vessel was overwhelmed by an explosion of seismic magnitude.

Like an inverse mushroom cloud, the explosion pushed down from the centre of the rig towards the shore, then rushed out in thousands of compacted mini-clouds of burning gas which projected 100 metres out to sea. In the control room of the *Tharos* Anthony Ashby, the deputy OIM, screamed at his crew to get down as the fireball punched at the reinforced glass windows, which were pushed in and bounced back out, but did not break. Fred Busby was just being helped into the *Tharos*'s hospital when crew rushed in with orders to 'secure all doorways'. On the deck of the *Tharos*, Powell could see 'a wall of flame black out the Piper'. On the *Maersk Cutter*, the crew on the bridge all ducked as the wide glass windows turned bright orange.

According to witnesses, the explosion seemed at first to drive the *Sandhaven* FRC down under the water before tossing it back out into the air, spinning in the flames. Only one of the twelve on-board would survive, the remainder either killed in the blast or (like the two other crewmen) drowned in the aftermath. Letham was blown out of the boat and into water now ringed with fire. There was also fire on Letham himself, and due to his life jacket and flotation suit he had to struggle to sink under the waves so as to properly douse himself. The heat of the fire above was burning his face and, although he did not know it at the time, the life jacket was beginning to melt onto his back. After recovering from the shock, he began to swim towards the leg at the south-west corner, but knew he could not stay there long – the heat was too intense – so he decided to swim upwind, between the risers and underneath the installation. When he looked up from his strokes all he could see were flames flowing across the bottom of the rig and radiating heat waves that stung his eyes, singed his lashes and burnt his face.

The men to whom Bob Ballantyne was heading, his friends, were blown away by the fireball, yet all he could think was: 'Thank God I didn't make it to them.' There was no time for remorse, terror was at his back and pushing him down off the cellar deck and into the water where, submerged up to his neck, he clung to the platform leg and began looking around. The sea in front was coated in oil, and elsewhere it was on fire; chunks of steel and bits of lifeboat were dropping down in front of him every few seconds, and through it all he could see was the retreat of the rescue craft. Although they were moving back to reposition, to Bob Ballantyne it felt as if he were being abandoned.

He began to shout: 'You bastards! You bastards! Come in and get me.' The blow-torch roar of the fire should have consumed his cries, but the FRC crews beyond the flames would later swear they heard him.

Those men on the immediate edge of the helideck were taken instantly by the flames, while those further back had time to turn and run. David Kinrade, whose radio messages had announced the abandonment of the rig less than 50 minutes before, finally made it off by springing across the helideck and launching himself into the air, believing his back to be on fire. Mark Reid would later describe the explosion as 'like an atom bomb' with a 'massive mushroom effect'. He began to run, but could feel a searing heat chasing him as he tried to reach the set of stairs on the east side, which would have taken him back to the helifuel tanks and the kitchen supply shed if he had not been beaten back by the flames. As he slowed to a walk at the centre of the deck, others were doing the same. Ringed by fire, there was nowhere to go. When a fireball suddenly burst over Reid, he quickly threw up his hands to protect his face and could feel the skin begin to burn, so he turned and ran towards the edge and threw himself off, while still remembering to hold down his life jacket by placing his arms across his chest.

The men had seen the flame before they heard the noise. In a split second the helideck, the entire rig, was girdled in flames, rising up hundreds of feet on all sides and touching at the apex, turning what lay within into a temple of heat. Meanen instinctively threw himself back and onto the roof, where other men landed on top of him. Instinct took over and, for some reason, he ran across the roof to where a steel radio tower rose 100 feet up. He squeezed through the metal safety frame that surrounded the ladder and began to climb the rungs. Then after a dozen or so steps he stopped and, in a moment of intense clarity, realized what a fool he had been climbing a ladder to nowhere. For the first time since the disaster began, Meanen thought he was going to die, yet this single thought proved to be the spark that ignited a burning will to live. In the next 30 seconds it was as if a machine had taken over his body. He backed down the steps, squeezed back through the bars and began to run across the helideck. When he reached the metal poles that stuck out and supported the safety net, he slowed down, took off his life jacket, stepped out on to the metal poles and looked down into the water 170 feet below. He then threw the jacket over, back-tracked, ran and jumped.

It was only when his back foot took off and his entire body had left Piper Alpha for the last time that, hanging in the air in that fraction of a second before gravity took grip, consciousness returned.

At that exact second, running through his mind was a single sentence: 'What the fuck have I done?'

He had six seconds to contemplate his actions, and as he fell he burned.

●

People who have survived falls from a great height have testified that the acceleration of gravity is so quick as to resemble being shot downwards from a cannon. Since an object such as the human body accelerates at roughly 20 miles per hour for every second it is in the air, after three seconds it's falling at 60 miles per hour. After six seconds the body reaches 120 miles per hour, or terminal velocity as it is known, where the wind resistance is equal to the force of gravity and the object can travel no faster. Joe Meanen and the others who leapt off the helideck reached terminal velocity a microsecond before they struck the water which, according to all conventional thinking at the time, would wipe them out. An object travelling at such speed does not give the water particles time to give and so they become – or so it was believed – as resilient as concrete at first contact. At the Robert Graham Institute in Aberdeen where oil men complete their offshore training, instructors advised that any leap into water from higher than 30 feet could be fatal – leaping almost six times this height was suicidal under any other circumstances.

Experts who study falls believe that the best way to land is feet first so that the balls of the feet make initial contact. However, this can still lead to hip-bones being violently jammed into the chest cavity and the destruction of internal organs. An alternative hypothesis (though one not applicable in this scenario) is that people 'roll' on impact in an attempt to distribute the impact across the body rather than concentrate it on a particular point. The mental attitude of the falling man is also a contributory factor in his survival; those who can remain relaxed fare better than those who tense their body which can lead to the force of impact shooting directly up the spine.

For three men it would indeed prove fatal. Alexander Duncan, aged 51, a steward with Kelvin Catering; his colleague William McGregor, 48, both from Aberdeen; and Michael Ryan, 23, a roustabout with Bawden International, were all found to have sustained chest injuries, fractures of the rib-cage combined with injury of varying degrees to the lungs, heart and liver, all consistent with a fall from height onto water.

Meanen fell through the air as if on a chair, one which at the last second tipped him to the left so that this side of his body took the impact. His T-shirt and track-suit bottoms were soaked from the spray of the *Tharos*, but his arms were bare and in those few seconds the radiating heat from the fire cooked both his hands and lower arms from the bottom of his sleeves down. Despite travelling at 120 miles per hour, he felt nothing as he struck the surface of the sea and sank approximately 20 feet below the surface, which he then felt he would never rise up and reach.

Upon surfacing he realized the current was moving away from the rig and he joined a slow-moving tide of debris and dead bodies. He knew then that he would survive. Though far from safe, he knew the worst was over; yet he felt nothing, no emotion, just the mechanical desire to retrieve his life jacket and reach an orange lifeboat he had spotted a few dozen yards distant. The boat, which had been blown off the rig in one of the previous explosions, now had its roof partly removed and was partially on fire. Meanen swam over to the side of the boat and began to scoop armfuls of water to extinguish the flames and then heaved himself in through the craft's side-entrance. It was only when he was sitting in the boat, trying to catch his breath, that he noticed his arms and the blisters that had risen up five inches, giving him the appearance of Popeye. At this point he could feel no pain.

Not everyone on the roof of the rig was forced to jump or die. Ian Fowler dived to the floor and put his hands on the back of his head for protection, feeling the hairs singe. Then he hauled himself up and, although there were men still on the ground, he scrambled over them to escape the heat. He managed to get across the deck and back down the stairs to the rear door of the galley, but the area was now completely filled with smoke to the point where he could not see the fingers on his hand. He began to feel his way along the walls until he reached what he believed was a small alcove that would lead to the reception door. He thought about going in, but reasoned that if he did so he wouldn't come out again. Instead, he set off for the food container.

In the water below, Ballantyne had stopped screaming when he noticed someone swimming towards him. He reached out and grabbed him in an act not of rescue, but out of a desperate need for companionship; whatever time he had left, he did not wish to spend alone. It was Ian Letham, and together they clung to the leg and watched as the east crane, hundreds of tons of iron and steel, tumbled down, sending salt water and sea spray over their faces.

As he clung to the leg, Ballantyne could see by the direction in which the debris was drifting that if he were to push off he would be carried away from the platform. Yet each time he thought of letting go, his fingers tightened on the ladder rung. For a few seconds the fear made him irrational enough to consider how best to get back on to the platform and away from the heat. The flames and incessant thrum of burning gas were so loud that, even with Letham right by his side, he had to communicate by gesture. He knew he had to let go and swim for it, but as the night had shown him, every action presented the possibility of dying. The water in front of him was coated in oil, and further ahead sat a long, thin island of fire. What, Ballantyne, thought, if the oily patch of water in which he was about to swim drifted into the flaming island and ignited? They would both be set ablaze. Yet he could see no option.

He gestured to Letham, who nodded, and together the two men pushed off.

The act of leaving the platform, of letting go of the rung, filled Ballantyne with an immediate sense of elation. He was convinced that he had made the right choice; he had faced his fear and would now turn his back on it. As he saw the oil slick drift closer to the flaming island, he flipped over onto his back and looked away. He later likened the decision to a child who hides from a monster by putting his hands over his eyes. As he kicked away from the platform and towards the burning slick, Ballantyne continued to shout and swear that the Piper would not kill him.

Drifting on a five-gallon drum a short distance away, Letham was also fuelled by rage at the loss of his colleagues, at his desertion by the other rescue craft. He shouted and kicked, moving further away with every kick.

At the very point when Ballantyne passed closest to the flames and the fear of igniting was at its peak, he found himself thinking of a trinity of reasons for living. He began to imagine Clyde Football Club finally lifting the European Cup; and then there was Pat, and the three-week holiday in France they had booked, the first they could afford. A parsimonious Scot, he might have been in a fight for his life but he was damned if he would lose his deposit. The strongest motivation came in the face of Amanda, the daughter he had not seen for six years since his divorce, and who he promised he would hold again if he was delivered from the sea.

Mark Reid also jumped 170 feet and surfaced, miraculously, without any further injury. The salt water was stinging his hands, but his immediate problem was the life jacket – it had become twisted and thus failed to balance the weight of his steel-toe-capped boots, which were now dragging him under. The waves washed over his face, causing him to splutter and struggle for breath, but once he had managed to kick off his boots he repositioned his life jacket relatively easily. Looking around, he saw a bearded man in some distress, spitting water and also struggling with his life jacket. Reid swam towards him, each hand stroke into the salty water making him wince with pain. He passed a lifebelt which read *MSV Tharos* and grabbed it. When he reached the man, he settled him down. Reid held on to the lifebelt, the bearded man held on to Reid, and together they drifted out into the night.

11. ADRIFT IN THE DARK

The *Silver Pit* was starboard on to the western edge of the rig when the MCP-01 riser exploded, wrapping those out on the exposed rescue area in a scalding blanket of heat. Ed Punchard was port-side and so furthest away at the time of the blast, but this meant he was last to reach cover behind the stairwell that led up to the bridge. The huddle of men was so deep that pressing against them still left him exposed to air hot enough to blister paint and smoulder rope.

Turning to look for another means of escape, Punchard saw Mahmood Khan (the first man rescued) clamber off the side of the ship. Peering over, he saw the technician in his boiler suit clinging to the rope netting, but protected from the heat by the ship's entire bulk. Punchard followed his lead, slipped through the open rescue gate to his right, grabbed a rope that hung next to the gate and lowered himself down to the point where his feet were just touching the water. The drop in temperature was an immediate relief.

Then the ship's engines started up, with the captain intent on putting distance between the vessel and the flames. While Khan was safe in the hammock of the netting, Punchard had tried but failed to reach the net and instead found himself being towed through the water, his heavy steel-toe-capped boots in place of water-skis. He began to shout for help, but it was a couple of minutes (or so it seemed) before George Carson, the *Pit*'s second engineer and acting medic, peered over the side, wearing a pair of orange ear-muffs. When he saw Punchard's strange pose his first response was to smile.

Punchard continued to shout: 'Stop the ship!' As more heads appeared to witness his plight, he was increasingly in danger of drowning. The strength in his arms was dwindling, he found it harder to hold on and began to slip down the rope, now no longer water-skiing but being dragged through the waves face-first. Each mouthful of water made him consider whether to just let go and swim for it, but he was frightened of being caught by the propeller.

The more exhausted he became, the more he tried to focus on the faces of Vicky and Suzie, his wife and daughter, and this helped him to hold on. The ratio of time spent above and below the water had reversed, and his vision was filled with green bubbles. There were only brief snatches to rise and suck in air. On one such crest of a wave, he saw the face of Gareth Parry-Davies leaning over the side and shouting: 'Fucking hang on! Fucking hang on!'

Punchard kept trying to kick off his rig boots, eventually succeeding with the right but the left refused to shift. Then he felt a sharp pain in his crotch. He had kept his legs wrapped around the rope, which hung in a loop, with the other end also fastened to the ship. In a rescue attempt the crew were lifting up the rope from behind, which served only to dunk his head further into the water as his backside rose. He was only saved when the engine was shut down and the ship eventually stopped, drenching him one final time with the bow wave. He was dragged aboard, limp and exhausted, by Andy Carroll and Parry-Davies. After a few minutes spent recovering his breath, Punchard realized that the experience had reinjected his body with another shot of adrenalin, the initial dose from his earlier escape having ebbed. During the next few hours he would experience what he later described as 'a great feeling' – a degree of clarity, focus and concentration that he had never encountered before and never would again. Despite being in the midst of a horrific scenario that involved smoke, fire, explosions and deafening bangs, he was acutely aware that this was the 'finest opportunity of his life' to make a difference. It was, he later said, 'a beautiful thing'.

During this time the *Silver Pit* had moved in an arc from the north-east round to the west side, where she now sat 400 metres off the north-west corner. Looking up at Piper, Punchard reflected that it was no longer on fire, it was being consumed; it seemed to him that the steel members on the cellar-deck were glowing red-hot while the water below was steaming. As he stood watching, Edward Amaira said: 'There'll be nothing left. Nobody left.'

Since the dive team's arrival on-board just over 40 minutes earlier, they had been operating as they had back on the rig, running ideas past Stan MacLeod as superintendent and Punchard as coordinator. As each new batch of survivors was carried back by James MacNeil and his crew, the dive team would help to haul them up and then down to the 'hospital' – as the one medical examination couch and a couple of chairs were called. When not below deck with the injured, the divers remained up top, scanning the waters for any survivors. Yet they had already realized that the *Silver Pit* was ill-equipped for its role as an emergency safety-vessel which would be ready at any time to take on board the rig's full complement of 220 men. The boat's bow thrusters had conked out five minutes after the first explosion, which had reduced manoeuvrability. When the gate at the side of the vessel was opened, it fell off its hinges. The internal layout, tight stairwells and sharp corners also made moving the injured difficult and painful. When they went to switch on the searchlight, they discovered there were no bulbs on-board. The crew were using an Aldis lamp, which had neither the same range nor the ability to spin 360 degrees.

Despite the dark, they could see the men on the wreckage with their naked eye, but Punchard and a couple of other divers couldn't work out if there were three or four men on what appeared to be a shattered lifeboat. They just knew they had to get to them fast. Punchard ran up to the front of the bridge and called to the captain to sail towards the wreckage, but he appeared not to hear him. Punchard repeated his request to the captain and eventually he responded. Sabourn was anxious not to clip the survivors with the ship's propellers, so he tried to position the vessel upwind of the survivors and then drift downwind towards them. Unfortunately, the vessel kept stopping downwind and so drifted past. When this happened a second and then a third time, the dive crew became increasingly frustrated.

Gareth Parry-Davies, who was still in his wetsuit, tied a length of rope to his waist and shouted: 'Tell the fucking idiot, I'm going to leap into the sea and get those guys out. Just make sure he knows what I'm doing.'

Punchard returned to the bridge and explained to Sabourn that a diver was going into the water, and that he himself would tell the captain exactly where the diver was so that he could manoeuvre the ship accordingly. He then leaned out of the window and said: 'Go for it, Gareth.'

Parry-Davies dived in and began swimming with his tether towards the wreckage. On the twisted, buckled and broken hunk of wood, Roy Carey managed to lift his head and wondered if his prayers had been answered.

Back on the bridge of the *Silver Pit*, Punchard was saying: 'OK. We've got a surface diver in the water. He's off the portside . . . He's ten feet away . . . He's fifteen feet away.'

Carey could see the diver draw ever closer, and with him the safety of the ship behind; then suddenly the diver seemed to stop in mid-stroke and move back.

Parry-Davies had been drawing nearer and nearer to the men when suddenly he felt his waist grow tight as the rope drew taut. The ship had begun to drift downwind once again and he was out of rope, a few feet shy. Recognizing the problem, Punchard shouted down to the rescue deck to 'find more rope and tie it on'. But the search was unsuccessful and Parry-Davies could only curse and splash the water with frustration before turning and swimming back.

The men were eventually picked up shortly before 11 p.m. by an FRC which manoeuvred the craft against the *Silver Pit*'s starboard side where the scramble net hung down as one final obstacle. The divers leaned over the side to assist the men in their climb, but they realized it was going to be more difficult than they imagined. The survivors were exhausted, freezing cold and badly burned, while one was quite petrified. Those wearing life jackets found that the bulk caught in the ropes. It was the fingers of one man that presented the biggest problem; once he had pushed himself up from the boat and taken a grip of the net, he would not let go, despite the pleadings and assurance of those around that they would not let him fall. Parry-Davies then climbed down beside him, whispered that everything was all right now and gently tried to pull a single finger off the rope, while the man wept and his body shook, his remaining fingers gripping all the tighter. It took minutes of quiet persuasion to get him up high enough and into the arms of the crew. Eric Brianchon wasn't going to let go of life.

The full extent of the men's injuries was apparent once on deck with their lifebelts removed. Punchard was startled to see that Roy Carey's head had been burnt down to the bone, his hair and several layers of dermal skin incinerated by the heat. Erland Grieve was also blackened by soot and smoke, with burns to his arms and face. Yet it was Eric Brianchon, the platform's only Frenchman (and a contractor with Coflexip) who was in the worst shape. He was practically naked, his clothes reduced to tattered strips; his body was pale, with patches blackened by fire, while all across it hung shards of fine skin.

Before Brianchon could be carried from the deck down into the hospital, it was necessary to move another injured man who was currently occupying the couch. Punchard helped to carry him to a nearby bunk before returning to collect the Frenchman. Although Punchard was assisted by another diver, Andy Carroll, it still took too many minutes, punctuated by gasps and shouts of pain, to manoeuvre Brianchon over the step of the watertight deck and round the bulkhead doors.

Brianchon was finally laid out on the couch with a blanket tucked round him. He was stiff and he lay with his knees bent up as Punchard tried to comfort him: 'You're all right now, Eric. Don't worry. You're safe.' But Brianchon replied only to say he felt cold.

With his thick spectacles, salt-and-pepper beard and wildly askew hair, George Carson liked to leaven even the tensest situation, joking that a third-degree burn to the face was 'Ah – just a wee while too long in the sun.' But he was frustrated. His designated role as medic was based on a two-day course in first aid, approved by the Department of Trade; yet even with this basic knowledge he knew that his supplies were meagre as well as being managed through a complicated system. The medicine was kept in bottles which were not named according to their contents, but numbered. The key to which number indicated what drug was kept separately in the captain's medical book. As he tended to Brianchon, Carson knew that burns of such severity required a saline drip to rehydrate the body as well as effective pain management. There were no saline drips on-board; there was morphine, but it was kept in a locked box with the captain who was the only one trained (back in 1969) in its administration. Since it was clearly impossible for him to leave the bridge, the drug was as good as thrown overboard. The only painkiller in Carson's possession was his own packet of paracetamol, left over from a previous trip when he had dislocated his shoulder, and which he was now using to treat a twisted ankle. He gave Brianchon two tablets with a glass of water and then went down to inspect the other two men.

Punchard was left alone with Brianchon, whose eyes were now fixed in a distant stare, as if transfixed by the fires he'd come through. They had met a few times before and Punchard knew he spoke little English but that once, in Barry Barber's office, they had had a stilted, tortuous conversation about the Grand Prix as well as Jaguar's recent success in winning the Le Mans 24-hour race. He tried to drag him into the present.

'Eric, do you remember we talked in the office about Alain Prost?'

Brianchon moved his eyes, looked up at Punchard and mumbled: 'Alain Prost.'

The act of drawing him back from the brink of unconsciousness had reconnected him to his pain, which heightened as his temperature began to rise under the warmth of the blankets. His face contorted and he cried out, while pulling the blankets ever tighter. Punchard found his agony difficult to watch and left as soon as he thought he could do no more, figuring there were others he could help on deck.

They then moved Roy Carey down to the hospital, where his wet clothes were removed and a blanket wrapped around him. Punchard was impressed by his spirit and stoicism. Whenever he enquired what he needed, Carey would reply: 'Don't worry about me. I'll be fine. No problems. Just leave me here. You look after the others now.'

As the hospital had room for just two men, Erland Grieve

was taken to a bunk in a cabin below. When Punchard looked in, both of them were anxious for a cigarette, but when he went off to speak to the crew about accessing some packs from the bond shop, they said that the captain had the only key and he couldn't be disturbed. This drove Punchard into a rage, in which he threatened to take the lock off with an axe for 'an injured man wants a smoke'. It could be argued that it was a sign of a swift recovery, to opt to fill one's lungs with a different type of smoke. It took a crewman reaching into his pockets for a packet of Benson and Hedges to calm everyone down.

•

The engine had never been right. Despite the close attention of John Faulkner, the *Loch Shuna*'s engineer, one of the outboard motors on the vessel's FRC had remained stubbornly temperamental. The *Loch Shuna* took a position one mile west of the Piper, and no sooner had craft and crew hit the water than the engine failed to start, an act of mechanical recalcitrance that had Ian Muir (the 27-year-old coxswain) cursing. Faulkner, who was crewing along with a third man, David Birnie, tinkered with the engine while in the distance Piper burned. Finally the engine kicked into life and they set off, the air now perfumed with petrol.

Less than thirty minutes later the craft could be seen floating a few hundred metres off the south-east corner. Illuminated by the flames, the faces of the crew appeared desolate and angry, yet the engine might just have saved their lives. They had first rendezvoused with the *Silver Pit*'s FRC roughly 800 metres west of the platform. The *Silver Pit*'s coxswain, James McNeill, had picked up a survivor but couldn't find his own ship, so Muir suggested he take the man – who had severe burns on the hands, face and head – to the *Loch Shuna*. He also proposed that both boats should work together. While McNeill and his crew departed, Muir steered the vessel around the *Tharos*, whose water cannon had started to discharge and he had no wish 'to be swamped by the deluge'.

When they arrived they cut the engine between 250 and 300 feet off the south-east corner, as close as the heat would allow. The options were poor. The platform was on fire from the 40ft-level up, and through the smoke they could see men waving, but as the platform overhung the legs the craft's route was blocked by a curtain of falling debris and burning liquid. They were faced with a difficult choice: 'We had a very quick discussion about the situation to consider whether to make a dash in and a high-speed pass to try to uplift people, but we decided it was not a very prudent move, given the state of the engines.' The decision now made, Muir began to steer the craft away. David Birnie looked back and saw more men climbing down to the platform; they were on fire.

It was the right decision. A few seconds later, with the collapse of the MCP-01 riser, the ball of gaseous flame pushed down to sea level and when Muir looked back the men were gone.

Standing over the steering wheel of the FRC, Muir was shocked by what he had witnessed and began to despair of anyone making it off alive. A few minutes later, while the men were in conversation about where to head next, there was a crackle of static on the radio. Muir answered and was told by the *Maersk Cutter* that survivors had been spotted in the water to the north as well as to the west of the platform. This raised their spirits, and with it the desire to do what they could not do for the men left behind to drift with the current.

The water had been fanning out from the cannon of the *Maersk Cutter* for over forty minutes, with no effect on a fire fuelled by its own supply of oil and gas. Yet the spray's high arch over the sea provided a cooler canopy under which Muir and his crew went to work. The craft slowed and moved forward at a crawl, the crew scanning the inky waters for any sign of life amid the clutter of chipboard panels, orange-boxes and chunks of wood. Then they spotted a man floating. He was still, with his eyes closed, and as they began to approach him a cry went out. Muir decided that the priority should be the voice in the darkness, which was coming from as close to the platform as the crew (whose faces were burning) dared go. The water was thick with oil, and all they could see was a blackness that shimmered with the current and glowed from the fire above. Then two white points appeared on the surface. The survivor, whose face was black with oil and smoke soot, had opened his eyes. It was Mark Reid.

Reaching down to grab him, they found that his survival suit had become waterlogged, swelling his weight; so to land him, wheezing, on the rubber floor was a struggle. A second man was also recovered nearby.

Everyone on board was struggling to breathe amid the smoke and the superheated air, so they retreated 300 yards and caught a few breaths before powering back towards the silent man. When the vessel approached, Muir could see that the man's mouth was biting on the neck of his life jacket. He figured he must have jumped from a great height, and that the impact had forced the jacket up and into his face and mouth. His neck appeared to be broken but he was still breathing, if only faintly.

A fourth man was also found nearby, suffering from bruising to his back and a similar shortness of breath. Now under the strain of seven occupants, the FRC's unstable engine stuttered and stopped again. It was a dangerous location to be reduced to half-power; the heavy stench of unburnt gas had begun to fill the air in the last few minutes and the condition of the survivors was weakening. Muir looked up through the fountain of spray at the platform, then peered beyond the flames that emerged from every part of the construction and saw the derrick topple slightly, correct itself and then concertina down and out of sight. As the craft moved back from the platform they passed the charred remains of a lifeboat, but there was no sign of life.

Once the craft had moved outside the water screen provided by the *Maersk Cutter*, they headed towards another vessel, the *Maersk Leader*, an anchor-handling vessel that was now taking survivors on-board. As there was another FRC unloading men when they arrived, Muir floated for a few minutes before moving into position. Due to the severity of the men's injuries only Reid was able to climb the ship's ladder, a feat which caused him considerable pain. Muir then followed Reid up the ladder and instructed the crew to ready the ship's store crane and then lift the FRC out of the water and on to the deck. The care with which the men were then eased out of the FRC was lost on the silent man with the broken neck who was found to have died.

The remaining three were led down to the ship's hospital, where they were stripped of their wet clothes and wrapped in thermal blankets. General first aid was administered as they took turns with an oxygen mask and bottle. One man was badly burned, but so cold that he was placed in a hypothermic bath in an attempt to raise his core temperature.

•

Oil production on the Claymore platform continued until 11.10 p.m. and when a shutdown began it was not an 'emergency' shutdown (which would have taken immediate effect) but a 'controlled' shutdown designed to prevent problems occurring with the Claymore's own compressors. Sandlin, the OIM, had finally ordered the shutdown at around 11 p.m. after he had managed to contact Occidental's Emergency Control Centre using a satellite link.

There is confusion as to when exactly Claymore first contacted the control centre. While Davidson insisted it was between 10.50 p.m. and 10.55 p.m., Occidental's head of telecommunications logged a time between 10.38 p.m. and 10.40 p.m. What is thought to have happened is that Sandlin first spoke to a Mr Bryce and then to J. L. MacAllan (Occidental's production and pipeline manager) between 10.50 p.m. and 10.55 p.m. While Sandlin was on the phone, his deputy Davidson received a further report from the *Tharos* that the platform had once again been enveloped by a massive explosion. He then shouted across the radio room to Sandlin to get him to ask MacAllan if they should shut down. Davidson believed Sandlin was asking MacAllan for advice. On the other hand, Sandlin felt the decision was his alone and, at that moment, realized that the situation on-board Piper Alpha was uncontrollable and a shutdown now vital.

The decision was about to be made for him. When MacAllan asked the OIM for the current position regarding Claymore's production and pipelines, he reacted with 'a certain degree of anger' – impotent frustration, as he later described it – to the news that Claymore was still on line. MacAllan immediately told Sandlin to shut down production, blow down the gas line to Piper and get in touch with Tartan using the VHF radio, then tell them to also shut down production and start blowing down the gas pipeline between Tartan and Piper. MacAllan then rang Texaco, who operated the Tartan platform, to emphasize the importance of getting the gas line to Piper blown down as quickly as possible. He also made arrangements for Flotta to depressurize the oil pipeline and for Total to blow down the gas line to MCP-01.

Prior to MacAllan's instructions, management on Claymore had not thought to contact Tartan, who were one step ahead having began to make preparations towards a controlled shutdown between 10.30 and 10.45 p.m. The wells were shut down in stages between 10.55 and 11.23 p.m. Tartan also decided against an 'emergency' shutdown in order to maintain full pressure in vessels and flow lines from their satellite fields.

Lord Cullen would later write:

> There was no physical reason why oil production could not have been shut down earlier on Claymore and Tartan. This would have caused an almost immediate reduction in the flow of oil that was fuelling the fire in the centre of the platform. As the fire on the 68ft-level was fed by an overflow of oil from the 84ft-level, any reduction might well have had a significant effect on the fire threatening the Tartan riser. If oil production had been shut down before 22.20 hours, this would probably have delayed the rupture of the Tartan riser. It is not possible to say that it would have prevented it ... Any delay in shutdown contributed to the amount of smoke and heat which was generated by the pool fires.

Yet the production of gas made little difference. In Lord Cullen's words again:

> As regards the depressurisation of the gas pipelines between Piper and Claymore and Tartan, it is clear that even if this had been undertaken at an earlier stage than it was it could not have had any material effect on the fire at Piper, having regard to the fact that the capacity of each platform to flare off gas was extremely small compared with the enormous quantity of gas contained within the length of pipeline in each case.

•

The *Tharos* remained in position despite the detonation of MCP-01 and the amount of debris the explosion had scattered across the vessel's lower decks, chunks of iron, steel and wooden boards. At the time, David Olley was on deck to oversee the arrival of the first casualties. The noise and radiant heat was so intense that the medic was mesmerized. With the strong feeling that the fire itself was a 'malevolent entity with a will of its own', he began to silently plead with this spirit that it would 'just stop for a moment to give us a chance to catch up'.

Olley explained:

We were carrying the first stretcher from the boat across the deck when an enormous explosion on the Piper produced a massive fireball which rolled across the water towards us, appearing to completely engulf the *Tharos* ... The heat produced was enough to blister some sections of paint. As the fireball hit, all we could do was to lay down the stretcher we were carrying and drop to the deck ourselves, covering our faces ... I would love to be able to say that my only thought was for my casualty but to be honest, I didn't think of it until later.

Meanwhile James Kondol, the *Tharos*'s deputy OIM, was the first to raise concerns about the danger of hydrogen sulphide (H_2S) clouds, a colourless toxic gas of which small amounts are to be found in crude oil.

At 11.07 p.m. Letty, the captain, spoke with the Coastguard for the first time, who officially appointed him as On-Scene Controller, a role which he had assumed from the first few minutes. His first task was to contact the *Loch Shuna* and appoint the captain in charge of the search-and-rescue operation, with the responsibility for coordinating the various ships that were already arriving on scene.

At 11.12 p.m. The *Silver Pit* contacted the *Tharos* to inform them that they now had 25 survivors on board, three of whom had serious burns and were in urgent need of medical attention. At almost exactly the same time a new message came from the *Tharos*'s helicopter, circling the scene, which reported that the entire platform was totally engulfed in flame.

At 11.13 p.m. Letty was notified that a Nimrod aircraft and four helicopters were already en route to the scene. As he acknowledged the message and clicked off the handset of the radio, he looked out across the scene and saw the drilling derrick collapse down onto the platform.

Yet there was no time to ponder the consequence for those still on-board. As he struggled to take in the view, messages came through either from Occidental or the Coastguard, ordering the *Tharos* to move back immediately because of the rising risk of H_2S and the possibility of an underwater explosion. After issuing the order to engage engines and move back 800 metres, Letty broadcast a message to all vessels in the area who were carrying survivors that they were to be brought to the *Tharos*.

At just after 11.15 p.m. the *Tharos* began its retreat from the Piper, the cooling water from the cannon of both the *Tharos* and the *Maersk Cutter* now landing in the sea. Piper Alpha and the men still on-board would enter the final act alone.

12. THE END OF ALL THINGS

He was dressed in a red survival suit, a bulky orange life jacket hung round his neck and a damp cloth pressed against his nose and mouth. Sitting on a chair in the fabrication shop, a compact steel workroom on the pipe deck, his face was blank, as if the anxiety and fear of the past hour had worn through his emotional circuitry which was now, blessedly, switched off. He could not move or speak; all he could do was breathe and, with the smoke in the air, even this seemed beyond him.

The man had been the only resident of the small steel shack – 10 metres by 6 metres and cluttered with steel pipes, elbow joints, tools and pieces of machinery – into which Ian Fowler and Harry Calder now ducked sometime around 11 p.m. The two men had met up outside the galley door after both escaped from the helideck and the MCP-01 riser fire, Calder leaping over the handrails, hanging on to a cable tray and then lowering himself down until he was outside the tea-shack. The pair had just tried to get to the side of the platform, but only got as far as the Aqua-Chem before being forced back by a sheet of flames. They figured that the access stairs to the pipe deck were too dangerous, so they lowered themselves through a gap in the flooring underneath the south stairs that led up to the helideck.

When they reached the pipe deck it was burning hot. They were forced by the surrounding flames to seek shelter in the fabrication shop, where over the next few minutes they were joined by other men. Although a respite from the intense heat outside, the room was still an oven; from the doorway flames could be seen outside, while inside the smoke had thickened to the point where some men were beginning to lie on the floor, desperate to suck in those first few inches of clear air into which the curtain of smoke never quite dropped, regardless of how thick it grew.

Suddenly Calder and Fowler heard a din of bangs and crashes. In the belief that structures were now starting to fall they took shelter (what little it offered) underneath a steel table welded to the floor in the centre of the room. There they pressed their cheeks to the floor and kept one another conscious by constant talking.

The clatter they heard had travelled a few dozen yards from a second steel fabrication shed, a drill store known to all on-board as the 'White House'. The walls were lined with heavy pieces of machinery, drillheads and pieces of piping, and men had began tossing these to the floor rather than wait to be struck. Approximately 20 to 30 men were estimated to be inside, a few of whom had retreated back into the 'House' when the water from the *Tharos* departed. Derek Hill, a crane operator, was among them:

> We felt this warm liquid coming onto us and the other chap and I thought it was oil that had come up. As soon as we felt it we ran back into the store. A lot of us carried on at that point and we came back and discovered it was water from the *Tharos*. It was coming down, it was cooling down and some fresh air was coming in with the water. I am not sure how long it lasted, it was not very long. There was another bang and that was the end of the water.

In the White House, as in the fabrication shop, men were lying on the floor. The building was surrounded by fire and smoke and rocked by the din of explosions. For the past twenty minutes John Menzies, 51, had tried to escape from either door, but without success. When Roy Thomson, just 24, arrived in the drill store, he and the men he was with had found protective fire jackets on the back wall and had put them on. Shortly afterwards a fire had broken out: 'Up in the top right-hand corner looking back into the tool store, there was a cable tray that came in. It was a small fire at the start in the far corner. It just started getting bigger and bigger and the fumes off it were choking us.'

The men, such as Andrew Mochan, were among the last to escape the accommodation modules. For almost an hour he had searched the corridors and staircases for a way out before finally discovering that the door adjacent to his office – previously thought impassable – was now clear, having been dampened down by water sprays from *Tharos*. Once on the pipe deck, he and those with him were herded into the White House by burning steelwork and debris. He said: 'It was complete darkness and in the first explosion I do remember there was a chap who landed on top of me, one of the Bawden chaps called Bob Patterson. The drilling bits were being thrown about like matchsticks. You were more or less huddled waiting to be hit.'

Sitting, also enduring the smoke, was William Lobban (28, a senior supervisor with Macnamee Service Ltd) and Stuart Sutherland (just 21, a student who had abandoned his studies and was working offshore as a cleaner with Macnamee). While trapped in the accommodation modules, Lobban had checked every exit from top to bottom and was heading back up when he bumped into Sutherland, panicking in a corridor so thick with smoke that they could barely recognize each other's faces. Lobban told him to grab hold of his survival suit and to stay close, and together they felt their way along B deck where they could smell fresh air which they thought was blowing through an air vent.

Then they heard someone shouting that he had found a way out. Lobban told him to keep shouting, that they would follow the voice which led them past the B deck offices, out of the door and on to the pipe deck, where they too briefly enjoyed the *Tharos*'s spray before entering the 'White House' in search of respite.

•

Hope had diminished along with the light and the breathable air in the fabrication shop where a dozen or so men were now preparing to die. A few sat on seats or wooden boxes, their faces black with smoke and each breath a loud, laboured wheeze. Panic had been burned off during the past hour or so, and what remained was a weary resignation. The knowledge that the end was near was carried by creaking groans and collapsing girders. 'You could hear the actual gratings, the walkways falling,' said Michael Jennings, who felt at the time, which was just before 11.15 p.m., that the platform was melting.

As if it were the stroke of midnight on Hogmanay, the men stood up and began to shake each other's hands, trying, as best they could in the smoke-filled darkness, to speak to everyone and saying: 'Goodbye' and 'Well, this is it' and 'This is the end.' A few men were crying. The weary, narcotic effect of acceptance and the loss of all hope settled over Jennings, who believed he was on the cusp of death: 'I was not quite sure which way it was [going to be], whether the smoke or the flames.'

And then, with the immediacy of vomit, he felt the over-whelming desire to live rise from his stomach up into his chest and, literally, raise him to his feet and towards the door. In the darkness he could not see where he was going and he tripped on somebody lying on the floor. He then tried to crouch down low so as to get away from the smoke, but even the deck was now burning hot and no place to lay a cheek. As the entire deck began to tilt, all Michael Jennings could think was: 'Good. It is going to collapse. At least it will be into the sea and it will be a lot cooler.'

In the darkness of the 'White House', Harry Calder was lying on the floor under the steel table, holding Ian Fowler's hand and trying not to cry. Both men were convinced that they were going to die. They hadn't expected to last this long, they had done everything they could think of to escape in the past 75 minutes, but it looked as if their luck had finally run out. Now they were trapped and terrified, and suffering so much smoke inhalation that Calder was vomiting onto the floor. The heat in the steel hut had risen above 120 degrees and the noise was a cacophonous soundtrack of roaring gas punctuated by sudden small explosions that still had the capacity to startle them, which was surprising given their current state of terror. During brief lulls both men could hear girders creak and groan and give way. Fowler muttered: 'I'm passing out ... I'm passing out.' Calder could only reply: 'C'mon, man.'

Then the whole structure tilted and suddenly dropped six feet. Fowler and Calder closed their eyes, tightened their grip on each other's hands and steeled themselves for extinction.

Weakened by heat, the steel support beams on which the platform had rested for thirteen years could now bear the load no longer. When they buckled and gave way, the result was like picking up a cardboard box with a heavy load and a weak base – almost everything fell through the floor. If a diver had been under the water at the time he would have seen units, compartments and modules fall from 68 feet, break the surface and then sink to the bottom.

Inside the perimeter mapped out by the steel jacket's four massive skirt pipes, a jumble of crushed steel boxes had piled up. What fell from below affected anything that sat on top of it.

At about 11.15 p.m. the western crane collapsed from its turret, with the jib and cab falling into the sea. A few minutes later there was a major structural collapse in the centre of the platform as B Module deteriorated; this caused the drilling derrick to collapse towards the north-west corner, with the top section falling across the pipe deck. The platform had already taken a slight tilt to the east, but it was about to drop dramatically to the west.

At approximately 11.20 p.m., the last of Piper's three gas risers blew up. The pipeline from Claymore to Piper Alpha contained 10 mmscf of gas and produced, according to Letty, the *Tharos*'s captain – 'an enormous explosion . . . the biggest of the night'. Once again Piper Alpha was hidden by a disc of white light, behind which the most dramatic collapse had begun.

Those men who were in the accommodation block were seconds away from sinking from view as, under the cover of a smoke-screen, the modules prepared to topple and then fall. The trapped men were divided between the minority, who attempted to escape an environment contorted into a hellish game of snakes-and-ladders, and the majority, who remained in the galley in a futile wait for rescue, only to be smothered in smoke. It was clear that a number tried to escape the accommodation block and failed, that in the confusion and the dark the corridors became a labyrinth that appeared to change by the minute. Exits previously impassable and pulsing with heat were cleared for a short window of opportunity by water from the *Tharos* or the *Maersk Cutter*, while those previously clear and accessible for the lucky few were covered in flames by the time others approached.

What kept so many men in the galley when there was no evidence of a rescue attempt but copious and compelling evidence of fatal danger, is tied to the psychology of the individual and his attitude to life offshore where there are strict rules, a chain of command and where men become conditioned to do as they are told. They were told by their OIM to stay and that help was coming – which it was, but it would arrive too late and even then could not penetrate the perimeter of flame. Colin Seaton did not issue an order that every man should attempt to escape the platform by whatever means necessary. Had he done so, as Lord Cullen would later conclude, more men might have survived, but he did not. He clung to the false hope of rescue and men died as a result. So did he.

Prior to death, those in the galley would have endured a rising level of fear, bursts of panic that overwhelms the body and the mind, as well as a brief flicker of hope. Others on the brink of death in sudden disasters have testified to the crushing sense of unfairness, disappointment almost, that this is it. The inhalation of smoke would have brought choking, a burning sensation in the lungs, the build-up of sooty deposits in the nose and throat, but each poisoned breath increased the level of carbon monoxide in the bloodstream which brought confusion, bewilderment and ultimately unconsciousness.

At 11.20 p.m. the Emergency Replacement Quarters contained at least 81 men. Among them were Alexander Taylor, 57, a roustabout with the Wood Group, known to his friends as 'Ronnie the Rocket' on account of always being in perpetual motion. He now lay still on D deck. On A deck lay Charles Duncan, 29, a floorman with Bawden International who had returned to work offshore after being made redundant as a barman. He was the father of one-year-old daughter Emma. David Henderson, 28, was also on D Deck. A lead floorman, he had been promoted before Christmas and so moved from the Claymore. His wife was due to give birth on 11 July. Shaun Glendinning, 24, a painter, lay on D Deck too, and had been married for seven weeks. Carl Busse, married not much longer, was also there, as was David 'Budweiser' Wiser.

They would have been lying on the floor, or slumped in

chairs or on tables. Autopsy results would confirm death by smoke inhalation, but could not rule out the possibility of men still being alive when the accommodation module entered the water. The atmosphere would have been black with smoke, but the fish in the tank may well still have been swimming.

Then the module tilted and the bodies began to slide across the floor.

The pipe deck had collapsed to the west at an angle of 45 degrees, splitting the structure from east to west along the line of the south face of the SPEE (Submersible Pump Electrical Equipment) module. It was this collapse that probably caused the Emergency Replacement Quarters to tip over and smash onto the Living Quarters West, whose temporary timber frame shattered under the weight. As it fell from the platform, it scattered huge chunks of board walls and floor into the sea.

As the Living Quarters West tumbled over, it was immediately followed by the Emergency Replacement Quarters. The four-storey steel structure fell from the platform, broke through the surface and began to sink, with each accommodation level taking in water until, by the time it crashed upside-down into the seabed 474 feet below, the entire quarters were flooded, with bodies beginning to float.

The seabed was to become a junkyard of debris. The giant steel pancake of the west helideck came to lean at a 45-degree angle against a leg. The exhaust ducts, each like a giant steel letter 'L', fell together, one landing face up and the other falling to the side. Inside the legs, steel boxes were piled on top of each other, rising up almost 100 feet.

In the 'White House', as the platform collapsed, men were sent crashing to the floor where they were clubbed by falling drill bits and pieces of pipe. A few men closest to the door appeared to just vanish, falling through the gap that suddenly broke open in the floor. Roy Thomson said:

The floor seemed to tilt a bit. I cannot remember in which direction ... I am not a good swimmer ... and I thought it was going to go into the water and it seemed to stop. After that there was another crash and the whole floor seemed to drop by six or eight feet. You could not see the pipe deck looking at the door. It was just black smoke and sparks.

The shelves collapsed and the men were hurled into a pile. John Menzies, 51, a scaffolder, had to wriggle out of his life jacket in order to free himself after shelves landed on him. Robert Paterson, 35, a welder, had both his feet trapped by drilling equipment and had to squeeze out of his trainers to escape. Derek Hill, 33, a crane operator, said:

The whole module fell to the west side. I do not know how many people altogether were in there, about a dozen and a half or something like that. A lot of people were just sitting around, when that tilted everything on the shelves and all the subs and things were flying about all over the place.

Once it had settled, Hill crawled through a metal frame that had slid across the front of the door. Thomson got his foot trapped in between pipes:

People were going by me out of the door. When I looked up the back wall ... was just a sheet of flames at the time. Once I put my leg out through the pipes there was a thing across the door. I climbed over it and the smoke [outside] seemed to have stopped. It was a weird kind of sight. The flames had all stopped and you could see the water going out to the west. The pipe deck at that time was all tilted down to the west side. It seemed to have split in the middle and the west side seemed to have dropped a bit. I saw people moving towards the west side across the pipe deck and I just followed them. I found the derrick had collapsed.

In the fabrication hut, everyone was knocked about too. 'We more or less landed in a heap in one corner,' said James Russell, 48, a mechanical technician, who heard a roaring noise and thought at first that they had tipped into the sea and the sound was gushing water. In the confusion it is not known what became of those who entered the huts and did not come out. They may have fallen through the gap that appeared or remained trapped under tools. When William Lobban got out, he realized that Stuart Sutherland never would.

There was no time to stop and search. Those who emerged from both the fabrication hut and the White House came out onto a mangled platform, but one where the visibility had improved. This may have been as a result of the accommodation blocks falling into the sea, carrying some fire and smoke with them, but no one noticed their absence. To get to the edge of the platform it was first necessary to jump over the gap where the deck had split, revealing a furnace below. The men then faced the obstacle of the toppled derrick which had landed between the White House and the west side, and each man would endure his own ordeal.

David Lambert, 39, whose first day on Piper Alpha was now coming to an end, was wearing dress shoes with thin leather soles and when he scrambled onto a gantry and began to edge along, his movements became slippery as his soles started to melt. He slipped, grabbed the handrail to steady himself and scalded his hand. Halfway along he and his group were engulfed in flames. The heat was unbearable and he knew that if he did not get off he would burn to death. So he took a running jump, put his hands over his head and jumped over. He had no idea what surface he would land on, but 150 feet down he struck the sea. His first thought was: 'That doesn't hurt', then he got a mouthful of salt water and realized where he was. Robert Elliot, who lost his trainers in his escape, sprinted barefoot across the scalding platform before leaping off.

John Menzies walked along the drill pipes, skirted onto a twelve-inch beam, and held onto a cable tray which allowed him to inch towards the edge of the platform. He looked out to see where the *Tharos* was and then leapt in, landing on his side with such pain that he thought he had had a heart attack.

In the 'White House', when Calder opened his eyes he could see a way out and hope coursed through him like a current. The floor and the shed had collapsed in such a way that a gap had appeared through which moonlight spilled in, and he and Fowler both squeezed out and onto the pipe deck. They were surrounded by flames, but saw someone climb up a rack of pipes and then disappear down the other side. Fowler shouted to Calder that this might be a way out and together they tackled the scalding climbing frame of the collapsed derrick, an effort that burned their hands and melted the soles of their shoes. On the other side, the steel beams rolled down to the platform's edge and as they slid down Fowler caught his foot, which stopped him going straight over the edge. He could see flames in the water below and, about a quarter of a mile out, a large semicircle of wreckage. As he looked at the wreckage he heard Calder come down behind him, but instead of stopping on the edge he hurtled by and over the side.

Fowler then noticed that he was not alone. There was another man on the edge, who advised staying on the platform rather than chance the fiery sea. Fowler looked down. He could see Calder resurface and wave to him to jump. Fowler did, curling himself into a ball as he fell.

When he surfaced his first thought was that Calder didn't have a life jacket, while he himself was dressed in both survival suit and life jacket. The pair swam towards each other and then began scanning the debris for a jacket for Calder, which fortunately floated by a few minutes later. The heat was fierce and they repeatedly ducked under to cool their heads. When they started to swim away, Fowler found his survival suit was filling with water, making each stroke exhausting. He grabbed a piece of debris, pulled himself up and soon fell asleep with exhaustion. Calder, after drifting beside him for some distance, was separated by debris and so they floated off in different directions.

When Michael Jennings emerged from the fabrication shed it was to his first deep breaths of relatively fresh air for over an hour. The smoke had been dispelled and he could clearly see the hulk of the *Tharos* 500 yards or so away. The environment where he now stood was a scrapyard of steaming white-hot metal; at the centre lay a towering pile of pipes over which men were scrambling before jumping off the platform and into the sea 120 feet below. 'That's for me,' he thought. 'I'm going.' But in order to reach the pyramid of pipes he had to first swing round a cable that was blocking his route. He slipped, put his hands down on the buckled deck for balance and screamed in pain as the skin on both hands sizzled on the steel. After reaching the pipes, Jennings decided that the easiest way to get down towards the platform's edge was to sit on the pipes and bump his way down. Yet halfway along the pipes began to separate, trapping his feet. He freed himself, thinking 'To hell with that!', then stood up and began to run down the pipes, hoping his speed and balance would get him to the bottom without taking a tumble. At the edge of the rig he stopped, looked down at the long drop and steeled himself to jump. Just as he was putting emergency training into practice, with one hand across his life jacket to hold it down and the other over his nose, he heard someone come up behind him, shouting that his feet were on fire. Before Jennings could react or move out of the way, he was pushed off the platform and fell sideways. A dozen or so feet down he struck a girder that sent him spinning for over 100 feet, a drop which was long enough to give him time to think that after escaping the platform he was now about to break his neck when he struck the water. Instead, his luck held and after resurfacing he lay back and enjoyed the surprising warmth of the water. He did not know if the sea's surface temperature had been raised by the fire, or if it was only a reaction to how hot he had been, but as he floated on his back he thought how nice it was to be away from it all. On a piece of wooden debris, with a training shoe as an oar, Michael Jennings then paddled off through a sea of bobbing duty-free cigarette cartons.

13. CHARGING THE FLAMES

The smoke trails had darkened the sky, creating an unnatural cloud cover through which the Nimrod MR2 flew. The RAF's principal maritime patrol aircraft had three main roles: anti-submarine warfare (ASW), anti-surface unit warfare (ASUW) and search and rescue (SAR). Capable of air-to-air refuelling, and with a flight range that could carry it north of Iceland or 4,000 km into the Atlantic, tonight's role was hardly taxing but certainly crucial. A crew of twelve, including two pilots, one flight engineer and an air electronics officer, would utilize the aircraft's sophisticated communication systems to act as a circling satellite, liaising between the growing number of vessels below, a swarm of helicopters and the 'beach' back home.

At 11.27 p.m. the Nimrod, call-sign Rescue 01, arrived on the scene and began to circle at a height of 30,000 feet. Shortly before arrival the crew had had the following exchange with the radio room of the *Tharos*:

> *Tharos*: '01, *Tharos*. The situation is that the platform is completely on fire from sea level to top. We have, in fact, pulled back somewhat. The structure is collapsing and it is total fire. We are continuing to spray water on it and we wish to evacuate non-essential personnel. We have taken and will be taking some casualties. Over.'
> *Rescue 01*: '*Tharos*, Rescue 01, copied.'
> *Tharos*: 'One of the standby vessels (*Silver Pit*) has reported

having twenty-five casualties which includes three serious burns and one injury. We wish to get rid of our non-essential staff so we can handle these casualties when we bring them on board.'

The Nimrod then sent an update to the Rescue Centre in Edinburgh (RCC):

Rescue 01: 'Edinburgh Rescue, this is 01. Sitrep. There is one surface vessel, semi-submersible (*Tharos*), three hundred yards on a bearing of 220 from the rig. He reports the rig is totally on fire from sea level to the very top. The structure is beginning to collapse. He is continuing to spray water. He believes that there are many casualties. He has seven survivors and another vessel in the area is believed to have twenty-five casualties on board: three with serious burns. This is Rescue 01.'

A few minutes behind Rescue 01 was the first Sea King SAR helicopter, R137, captained by Flight Lieutenant Stephen Hodgson. During the hour-long flight the cockpit window had been open, but as the helicopter reached a perimeter of just over one mile from the platform the heat waves were so intense that the crew closed the window as a protective barrier. It was decided not to proceed any closer than a mile upwind of the fire.

Rescue 137: '*Tharos*, this is Rescue 137. I'm in the area looking round. There's not a lot I can do in terms of the burning rig. Are you the rig that currently has a helicopter approximately two miles north-west of you?'

Tharos: 'Rescue 137, that is probably Yankee Bravo, our S-76. I believe you are the yellow rescue helicopter that passed my bow five minutes ago.'

Rescue 137: 'Roger. I've just turned round and put my lights on. I think Yankee Bravo has departed the area. If you've got casualties I can probably get to you, but I cannot get any closer to the rig because of the heat. How about the seriously injured you mentioned? Do you have the facilities to look after them or do they need moving first? Over.'

Tharos: 'We have an intensive-care hospital. The casualties are on one of the smaller boats at this time. We wish to lift them to *Tharos* and commence initial treatment.'

The Sea King Rescue 137 touched down on the helideck of the *Tharos* at 11.30 p.m., shielded from the heat by the vessel's water spray. The immediate priority was to dispatch non-essential personnel from the *Tharos* to the Claymore and then to return with the platform's medical personnel. Once this was completed, Rescue 137 was dispatched to perform searchlight sweeps of an area north and south-west of Piper, where it was most likely that the wind would carry survivors.

The second Sea King Rescue R117 arrived at 11.44 p.m., followed four minutes later by R138, on board which Paul Berriff, crouching behind the pilot, had already begun to film.

The view from the cockpit, while still a few miles out and at a height of 200 feet, was staggering. The sky on the edge of the horizon was still the light blue of late summer, but above the platform, and stretching out on all sides, it was as black as tar. Immediately below the Sea King the water, in a thick band, had turned a golden-orange colour, and ran like a broad runway towards the heart of the inferno. Piper Alpha was almost completely lost behind a cloud of brilliant white light, whose crown appeared to be 1,000 feet above sea level; only a small dark shape close to the waterline and resembling a crooked letter 'T' was evidence of a man-made structure.

The faint shadow of the *Silver Pit* could be seen a few hundred feet back, while the futile arcs of the *Maersk Cutter*'s sprays were visible landing in the water far back from the flames. The *Tharos*, illuminated with white lights, like a small city at sea, was 700 metres back, its aluminium bridge raised at a 45-degree angle, while the various water jets formed a protective cloud of evaporating droplets.

'Bloody hell, it's really on fire, isn't it?' said John Dean, captain of 138.

'It's a big one,' agreed Pat Thirkell, the radar and winch operator.

'I'm sweating already,' said Bob Pountney, the winchman.

In the water, a 4-foot swell had developed which gently raised up and dropped down a scattering of yellow plastic hard hats, charred pieces of tables, broken chairs, shattered wardrobes, pieces of paper, foolscap as well as official forms, bulk packs of cigarettes, boxes of King Edward cigars, chocolate bars and training shoes, dozens of plastic storage bins and the occasional paperback novel. In amongst the refuse were the men, clinging to bits of wood, burnt and broken lifeboats or bobbing, exhausted, in their bright orange life jackets. The dead lay face-down, carried by the current.

The *Silver Pit* was picking up whoever she could, with divers positioned on port, starboard, bow and stern, all scanning the waters for signs of life. Andy Carroll found the job frustrating; while some men were being picked up, others briefly glimpsed were being lost in the dark. He had spotted a few men waving frantically and had waved back to them, but then they were gone. 'It was heartbreaking,' he told Ed Punchard. 'It's insane to have survived the fire and then to be lost in the sea.'

Carroll urged Punchard to speak to the skipper and get more boats into the area, but when Punchard reached the bridge he found the captain preoccupied. He resisted the urge to shout and, instead, calmly said that they needed to get all nearby ships to launch their lifeboats. 'There are men in the sea. I'll operate the radio. Is that OK?'

The captain nodded and Punchard put out a statement:

This is *Silver Pit* calling all ships in the vicinity. There are many survivors floating on the surface between *Silver Pit* and *Tharos*. That is between the north-west area and the south-west area of the platform. Any ships with Zodiacs, lifeboats or small craft of any description, launch them and come to this area. There are many survivors in the water.

The HT24, the *Silver Pit*'s fast rescue craft, was blackened with smoke; the plastic sponsons, the sausage-shaped flotation devices that girdled the boat – were partially melted and bashed from where the coxswain had repeatedly wedged the boat against the rig legs to provide a stable platform for the rescue of desperate men. More than an hour had passed since the 'easy' evacuations of the north-west corner, and since then James McNeill and his crew had skirted the flames, dodged debris and narrowly escaped decapitation when a metal spike, hurled with the accuracy of a javelin at the head of Charles Haffey, had missed by millimetres. Haffey was not immune to injury, as debris from a burning hosepipe landed on his bottom lip, painful but preferable to being smacked by a wooden plank (another obstacle he dodged with a well-timed duck).

The heat under the platform, where the flames rippled like a reflection of the sea below, baked their faces and prompted McNeill to roll up the sleeves of his survival suit, no easy task given the tight seals. Yet the crew's success as 'fishers of men' was remarkable. For almost ninety minutes they plucked men from the waters and landed them on whatever supply vessel was nearest to hand, usually the *Silver Pit*. They looked up as one man leapt from the helideck, spotted where he landed and were in place with outstretched hands as soon as he surfaced from his long submersion. Among the most disturbing sights was a survivor floating on a piece of debris wearing only the elasticated trim round what had been the waistband of his underpants, which along with every other stitch had been blown or burned off. He was hauled on-board and McNeill saw that the skin from his arm and hand had peeled off and was dangling down like a lady's evening glove. 'Don't let him flake out,' McNeill had shouted to Haffey. 'Slap his face, slap his face.'

When rescued men were desperately cold they were placed up against the craft's engines and repeatedly questioned just to keep them awake. Haffey spotted one of the luckiest survivors, a man who had already sunk down a couple of times and had submerged once more when the craft reached the spot, allowing Haffey to reach under the water and haul him up by his hair.

During the night doubt never once washed over McNeill, who was driven by adrenalin and at the wheel of a craft he would later describe as 'the most beautiful boat'. He felt unstoppable. As the HT24 was driven by water jets and had no outboard engine or propeller to snag on debris, he was able to skim over large chunks of material by throttling down and utilizing them as aquatic ramps.

At one point, as they prepared to go into the platform on either the third or fourth rescue run, Charlies Haffey shouted above the fire's roar: 'I hope God is looking after us tonight.'

McNeill shouted back: 'Two good men are looking after you tonight, Charlie. Him and me.'

At one point, possibly after 11.30 p.m., McNeill's luck simultaneously gave out and yet held tight. The boat had just picked up three more survivors and was moving about 100 yards off the platform when he felt a strong underwater rumble. He shouted at the crew to get down and lie with their full weight on top of the survivors to protect them, and as they did so the entire craft was thrown up into the air. It appears that they had been sailing over a gas pipeline which had cracked, firing a bubble of high-pressured gas up and under the boat and splitting the hull beneath McNeill's feet. The pressurized gas shot up through the split and into McNeill's face, singeing off his lower eyelashes and burning the bags under his eyes, a sensation he described as like 'having hot sand thrown in your face'.

A gas bubble expanding up through the sea displaces the water, reducing buoyancy and forming a hole into which a boat will drop. Why the craft didn't sink, McNeill never knew, but instead he managed to force the throttle and the boat skipped on.

The engine, while still running, was losing power, and the sea was beginning to seep through the splits in the hull and pool around McNeill's boots; in a few minutes it had already risen above his ankles. The boat had come to rest about 60 feet from the platform, and the crew could see the state of collapse. On previous rescue trips the fire had run along the underside of the platform, but this underside had collapsed, sinking down into the water, the high empty space replaced by a tangle of twisted metal. When McNeill looked at the twin flares he thought he could see the metal melt and roll down the steel arms like lava. Then they spotted a hand.

The crew glimpsed, under the platform and partially obscured by smoke, what they believed was a hand, waving for attention. The man was roughly 100 feet under the collapsing platform. It was a difficult and dangerous manoeuvre to undertake even if the craft was fully operational, and the odds dropped for a boat that was leaking and losing power. To reach the hand meant speeding round flaming debris and between collapsing pipes and then, once he was on-board, they would have to navigate the short corridor still clear of collapsed structure until they shot out on the other side of the platform.

McNeill turned to his men: 'Well – are you willing to go? Are we going for him? We're his only hope.'

The reply came as a chorus: 'Go. Go. Go for it!'

McNeill gunned the engine and told his crew just to grab the man's hand and tow him out if necessary, as once in there was no way he was stopping. As the survivors on-board huddled down on the craft's bows, Haffey, Kiloh and Clark positioned themselves on the edge with their arms outstretched. When they entered the perimeter of the platform, the heat intensified and the boat rocked from side to side as it dodged debris. McNeill never looked up, only straight ahead. As they approached the man three sets of arms stretched out and snatched him up as they passed by. Before McNeill had even cleared the other side, the survivor was lying on the deck. His face was painted grey with smoke residue and he was coughing and spluttering, but he was alive.

The survivors were deposited on supply vessels. Then, in an attempt to bend the laws of physics and fluid dynamics with optimism and raw courage, the crew set out yet again to scour the flaming sea, although they did not get far. The water continued to rise and while the engine struggled on, churning away even when underwater, neither it nor the crew could compete with gravity and weight. McNeill radioed the *Silver Pit* to say they were sinking and that there was nothing else they could do. The water collecting around him at the stern drove the craft down and under, raising the bow up into the air where, supported by the remaining sponsons, it bobbed like an arrowhead. Forced to take to the sea, the three crewmen clung to the sponsons around the bow, while McNeill lay back and relaxed, the water was not too cold and he floated a few feet away, buoyant on a wave of pride in his men and deep satisfaction that he had done all he could.

In under two hours, four men in one boat had saved the lives of over thirty others, many whose names they would never know. Yet now the rescuers were in need of rescue themselves.

•

The time was a few minutes before midnight, less than two hours since the first spark. In the communications room of the *Tharos*, David Robinson was directing radio traffic while James Kondol went down to the room rigged up for receiving casualties. It had grey walls and was the size of a small tennis court, with brown blankets laid on the floor for each new arrival. As yet there were only a few men on-board, and, while crew served soup and tea from huge steel pots into white polystyrene cups, he listened as the survivors spoke of gathering in the galley and the lack of firefighting equipment. Afterwards Kondol went back up on deck and watched the platform's steady collapse. During the destruction the gas risers, as well as the main oil line, had been sheared off just above the waterline with an effect that his boss, Letty, described as like a 'Bunsen burner' emitting from the surface of the sea.

A message had just gone out to *Tharos*'s helicopter: 'Tharos/ Yankee Bravo. As you have already been told the risers are obviously broken and the oil is coming straight up onto the sea and burning as well.'

In the distance the *Loch Shuna* was struggling to cope with the coordination of the surface search-and-rescue, and in forty minutes' time would request to be relieved of the role as the captain tried to look after an injured man on the bridge as well as deal with the expanded VHF radio traffic. The *Lowland Cavalier*, to whom the task would be passed, was 800 metres north-west of the platform. The captain, Michael Clegg, had already ordered the vessel's own lifeboat to join the search and was becoming frustrated by the number of times sightings of survivors in orange life jackets turned out instead to be orange debris, buoys, fire boxes, and even a fish box.

While the *Maersk Cutter*, having moved back after reports of H$_2$S gas, continued the search, its FRC piloted by Ian MacKay drifted and scanned the waves, eventually hauling aboard one body but no new survivors.

The *Maersk Logger* was more fortunate, spotting Vincent Swales:

> The sea was only a four-feet swell, but when you are a man alone in the water you feel a little isolated and the boats look a lot, lot bigger. It was difficult to swim against it. Every so often the waves would break over you to keep you cool. Every time you were on top of a wave, you would wave, and then when you were down in it, you would swim to try and make your way over.

The *Maersk Logger* started moving towards Swales, but he had no idea whether or not they had seen him or were responding to another sighting, and that he would go under the propellers. Only when a rope was thrown down did he relax. The *Logger* also picked up Steven Rae, who – once he had changed into dry clothes and was wrapped in a blanket – made his way up to the bridge to ask the captain if he could call his mother. The captain told him that the radio was in use just then, but he would see what he could do a little later. And he was as good as his word, for within an hour Rae had woken his mother and told her: 'I'm fine – there's been a terrible accident, but I'm OK. I'm OK. Don't worry. But I can't talk now.'

Bob Ballantyne had drifted far from the platform. His face and hair were black with oil when he spotted the lights of the *Lowland Cavalier* shortly before midnight, and began to wave with what little energy he had left. It was enough to attract the crew, who threw him a lifeline. He grabbed hold but could not pull himself up. Three times he tried, only to bounce off the boat and back into the water. What strength he had left was not enough to raise his weight up the scramble net, and for the first time he contemplated giving up. He thought: 'I'm just going to let this rope go. I can't take it. I'm just going to drift out to sea. I don't care any longer.' Recognizing his distress, the Norwegian seamen gestured to him to tie the rope around himself. Three seamen then dragged him up the side of the ship and on-board. As Ballantyne could no longer feel his legs, he believed he must have injured them very badly, and he shouted: 'I've not got any feeling from my legs down – I've no feeling from my waist. Something has happened to me! Something has happened to me!'

On board, a more accurate diagnosis of his condition was carried out. Apparently he had failed to close his survival suit properly and it was filled with water from the waist down, rendering him artificially weighted and so unable to move his legs. He then lay on the deck 'like a beached whale' while the water drained out.

Afterwards he was taken below deck, provided with a change of clothes and advised to shower in an attempt to warm himself after such an extended immersion in the North Sea. However, such was the heat he had already endured that Ballantyne could not tolerate the feel of hot water.

'It was just burning up on my face and body. I couldn't bear the hot water. I had to put on the cold water.'

As the *Lowland Cavalier* had already picked up about a dozen people, the ship's medical supplies were low, so the only respite from the burning sensation on his face was by the application of a frozen packet of Batchelor's peas. Ballantyne found himself exceedingly emotional, moved by the simplest gestures – a cup of tea, or someone sitting and talking to him. As the night moved into the early hours of the morning he found it increasingly difficult to hear, but at one point he did hear a radio transmission that revealed the number of people rescued. He could not believe it was so low.

•

The number of burns victims on-board the *Silver Pit* was escalating beyond George Carson's ability to cope. Even getting them aboard at all was becoming more difficult. By midnight there were around 30 men on-board. Those men found drifting in the water were often weighed down by survival suits which, improperly zipped, had become waterlogged. The cries of those with burned hands as they grasped the rough rope netting was as distressing as their appearance, covered in smoke dust that coloured them ashen grey.

As Stan MacLeod described:

> People with very badly burned hands, it was obviously agony for them to climb up the scramble net, but some of these people were able to help themselves to a very great degree. There were other less fortunate people who had severe body burns and were really only semi-conscious. We had a phenomenal amount of difficulty recovering them, and we obviously inflicted a lot of pain and suffering on these people while we pulled them out of the water.

The seriously injured were brought to the cabin, while the more able were guided to the galley. Among Carson's concerns was that the injured would pass out, which could prove fatal, so he paired up injured with uninjured with instructions to keep them talking. However, this further restricted the available manpower.

There are differing accounts of how the most seriously injured men came to be removed by helicopter from the *Silver*

Pit. Walter Mitchison remembers George Carson telling him: 'There are a couple of blokes down there who want proper medical doctor's assistance, you'd better get a helicopter, Walter.' He replied: 'OK,' then went to the bridge where the captain said: 'Away you go, carry on. I'm watching the ship and you carry on and do as you wish.' When Mitchison then radioed *Tharos* and explained that they had badly burned men who required hospitalization, the radio operator replied: 'I'll be with you in five minutes.'

Ed Punchard insisted that the procedure was more problematic. After he spoke to George Carson, priority was given to Eric Brianchon, who was still drifting in and out of consciousness; Roy Carey, who masked bad burns behind a cheery countenance; and Erland Grieve, whose hands and face were severely burned. They were the top three on a list of a dozen injured men that Punchard handed to Stan MacLeod, who was on the bridge and who then tried to raise *Tharos* on two VHF channels – 16 and 8 – only to find them jammed with traffic. When a break emerged, MacLeod leapt in and instructed *Tharos* that *Silver Pit* had three seriously injured survivors who required immediate evacuation. The men had burns and lung problems, and he insisted on 'priority evacuation by helicopter'. He also offered a list of all survivors on-board. *Tharos* promised to get back to them.

Punchard and MacLeod then began to plan for the helicopter's arrival. Both wanted the men lifted off the large rescue deck, but the captain was adamant that they be carried on stretchers to the back deck where a large yellow circle marked the official spot for helicopter evacuations. The captain must have had good reason as, given the ship's layout, Punchard felt this put extra strain on the injured, but he set about finding stretchers; he was disappointed to discover only two, but figured the helicopter could provide others. Eric was strapped in first and as Punchard and the other men struggled to manoeuvre him up stairs and round tight corners, he moaned in pain. Punchard felt it was like moving a man in a coffin.

Rescue 138 touched down on *Tharos* at 12.07. Berriff's camera continued to roll as the four medics jumped out, one carrying a large cardboard box of supplies, and ran across the helideck. On the flight back from *Ocean Victory*, the crew had been disturbed by the extent of the platform's continuing disintegration and the ferocity of the flames. 'My God, she's really boiling,' said one crewman.

Around this time the Nimrod, Rescue 01, was updated by the coastguard on the numbers on board.

> *Rescue 01:* 'All stations. The number of persons believed to be on board the Piper Alpha oil rig was Two Two Zero. I say again. Two Two Zero. Intend searching around oil rig using surface vessels and helos. This is Rescue 01.'

As Rescue 138 waited on the helideck, the captain received news of his next task.

> *Tharos:* 'Rescue 138. This is *Tharos*. We have no passengers at the moment. If you would like to stand by. I have made contact with the vessel *Silver Pit*, who now has some 30 casualties. I believe 35, including some injured. I was going to contact Rescue 01 to check the feasibility of commencing winch recovery of those people to me.'
>
> *Rescue 138:* '*Tharos*. Rescue 138. Roger. Do you have position of *Silver Pit*?'
>
> *Tharos:* 'He said he was half a mile east of the platform. He's moving closer to *Tharos* and is listening on this frequency. He's a converted fishing vessel with a small H landing area. No helideck.'
>
> *Rescue 138:* 'Rescue 01. We are now actually on the helideck

> of *Tharos* and lifting to go to *Silver Pit* to investigate
> winching possible casualties to *Tharos*.'
>
> *Silver Pit:* 'Rescue 01. We have three seriously burned men
> and they will be in rescue stretchers. Over.'
>
> *Rescue 01:* 'Roger that. We have Rescue 138 closing your
> position and Rescue 117 to follow for winching oper-
> ations to remove survivors to *Tharos*.'
>
> *Rescue 138:* 'Rescue 01 and *Silver Pit*. Due to the large
> number of vessels in the vicinity we are having difficulty
> in locating *Silver Pit*. Can he illuminate himself or fire
> some type of flare?'

In the body of the Sea King, Paul Berriff checked his camera. He was anxious to film the recovery without impeding the crews' work (a skill he had refined over the past 41 sorties since filming began on 1 April), but he was also concerned about the need to ration film. Twenty-five minutes wasn't much, and it was going to be a long night. He moved aside to let Bob Pountney, the winchman, pass. Pountney was dressed for sea rescue in a yellow one-piece immersion suit with black integral boots, with a standard-issue life jacket that contained flares and a personal locator beacon in case he had to ditch. He was also wearing the blue bosun's-chair harness that would be attached to the wire winch. There were two types of stretchers by the door: the small Neil Robertson stretcher consisted of a long wooden board, with a cushioned head-rest and a heavy canvas cover that encased the body once the straps were secured; the Stokes Litter stretcher resembled a cross between a cradle and a coffin, with an orange-cushioned base and a perimeter of steel bars. One could fit comfortably inside the other.

The Sea King's searchlight picked up the *Silver Pit* and began to drop to the maximum safe height for a sea rescue, between 40 and 50 feet. Pat Thirkell, the radar operator, moved forward to act as winchman, a role which gave him de facto control of the entire aircraft, not just the rise and fall of the cable. The captain was in the cockpit 20 feet up in front. When the craft was in winch mode, every instruction in all three planes – height; left and right; forward and back – was given by Thirkell.

Rescue 138 used its spotlight to pick out the *Silver Pit*, which was resting almost a mile back from the platform, bow-on to the blaze. The craft began to hover and Thirkell slid open the heavy steel door, while Pountney hooked himself up and leaned out. The stretchers would follow him down. From now until Pountney and the men they had come to fetch were safely back on board, Thirkell would keep up an unrelenting description of every action on the line, painting a word picture for the pilot and issuing him with necessary instructions and positioning.

On the yellow spot on the *Silver Pit*'s stern stood Ed Punchard, who looked up as Pountney swayed down and bounced off the handrail before coming to rest on his knees, where he pulled at the cable for slack. The thrum of the rotor blades meant that Punchard had to shout at Pountney, to tell him Eric Brianchon would be on deck in a minute. When he arrived under the arms of the other divers, he was barely conscious. Pountney secured him into the stretcher and together they began to ascend.

Punchard watched, then said: 'Bon voyage, Eric. Good luck.' Fifty feet above, Berriff filmed the ascent by leaning out of the main cabin door. The footage shows Brianchon with his hands bound up with white bandages like cotton candy, swaying gently above the waves while Pountney leans over the stretcher offering words of reassurance. On the deck below, his colleagues stand looking up. The soundtrack has Thirkell explain in a mildly narcotic voice: 'Steady ... steady ... steady ... he's moving gently towards the fire ... little bit of swing.'

During the operation the *Silver Pit* had killed her engines, leaving her at the mercy of the current and making positioning more difficult for the pilot:

> *Rescue 138:* 'Silver Pit. Silver Pit. This is Rescue 138.'
> *Silver Pit:* 'Silver Pit.'
> *Rescue 138:* 'With your engine stopped the roll has made it quite difficult for us to get stretchers off. Can you get under way, sir?'
> *Silver Pit:* 'Yes. It will be very slowly as at the moment we are having a few problems with manoeuvring. Over.'

The procedure was repeated twice more to collect Roy Carey and Erland Grieve. When Berriff filmed Carey's ascent, Punchard spotted the lens and pointed it out. 'That's astonishing. What the hell is that doing here?'

With the three men on board and Pountney off the winch wire, Thirkell said: 'Lovely. That's all on board. Up, up and away.'

At 12.48 a.m. Rescue 138 touched down on *Tharos*.

At 30,000 feet, flying in a precise figure-of-eight (a movement specifically designed to benefit the on-board radar) Nimrod Rescue 01 continued to send situation reports back to the mainland:

> *Rescue 01:* 'Edinburgh Rescue from Rescue 01 ... Rescue helicopter 137 is searching an area to the west of the rig; Rescue 131 is investigating survivors in the water to the north-east of the rig. Helicopters Bristow Five Zero Yankee, Tango India Golf Bravo, Tango India Golf Oscar and Yankee Bravo (*Tharos*'s S-76) are en route to the scene or holding on oil rigs. They will take over the search when other helicopters have to refuel. The rig is now believed to be emitting hydrogen sulphide and

there is a high probability of an underwater explosion. There is heavy smoke and flames spreading to the north of the rig. This is Rescue 01.'

An attempt by Rescue 01, under pressure from the mainland to discover the number of casualties, was unsuccessful:

> *Rescue 01:* 'Rescue 01 to *Tharos*. Request the number of survivors and casualties you have accounted for so far. Over.'
> *Tharos:* 'I appreciate your request for casualty numbers but we are unable to pass them at this time.'

After midnight survivors were transferred on to *Tharos* using a rope basket, which had a broad circular base and rope sides that curved round into the shape of an Indian teepee. It was lowered down onto the deck of each supply ship or standby vessel that came alongside, and those who were able-bodied climbed in. During normal usage, it was forbidden for men to go into the centre of the disc – they were expected to cling to the sides – but given the ordeal they'd gone through and their distressed and frightened condition, they were ushered into the centre where they sat down. Meanwhile, four men from the *Tharos* escorted them by forming a protective ring around the sides.

At 12.40 a.m., Letty, on the *Tharos*'s bridge, decided to pull back far enough to be able to turn off the heat shield. There was little of Piper Alpha left and he felt it was more important to allow the helicopters to land and take off as easily as possible. Shortly afterwards he went downstairs to the hospital where, for the moment, medics from *Tharos* and neighbouring platforms were supported by those members of the diving crew with medical training. A few minutes later, at 12.58 a.m., a helicopter touched down with a team of five doctors from Aberdeen Industrial Doctors, led by Dr Ronnie Strachan.

The consequence of exposure to fire are smoke inhalation and burns. While mild smoke inhalation will often cause no serious lasting effects, if the smoke is hot enough (or exposure sufficiently prolonged) the airway and lungs will be burned. This is caused by both the temperature of the smoke and the noxious chemicals contained in smoke. These chemicals can dissolve in lung fluid, becoming corrosive. The upper and lower airways then swell, restricting airflow to the lungs. As a result each survivor upon arrival on *Tharos* was checked for soot in the mouth and nose, singed nasal hairs and burns to the mouth, as these are useful indicators of smoke inhalation. Those wheezing badly were provided with an oxygen mask to assist their breathing.

Burns, meanwhile, are characterized as superficial, partial thickness and full thickness, in ascending degrees of seriousness. Superficial burns damage only the outer layer of skin (the epidermis) and while they may become red, moist, swollen and painful, they whiten or blanch when touched and do not develop blisters. Partial-thickness burns damage both the epidermis and the inner layer (the dermis) and will develop blisters that ooze a clear fluid which has leaked into the burns from blood vessels, causing swelling and pain. While superficial and partial-thickness burns will usually heal themselves within a few weeks, depending on severity, full-thickness burns require skin grafts.

Full-thickness burns occur when the heat or flame has been so extensive as to penetrate through the epidermis, the dermis and down into the underlying layers of fat, destroying in the process sweat glands, hair follicles and nerve endings. While the most serious, this often renders them painless to the touch. The skin becomes leathery and takes on hues of white, black or bright red. The complications associated with extensive full-thickness burns include hypovolaemic shock as the fluid seeps from the victim's blood to tend to the burned area, which in turn can lead to a state of shock where the loss of fluids causes the blood pressure to drop so low that not enough is reaching the brain and the major organs, causing weakness, fainting, nausea and vomiting. Deep full-thickness burns can also destroy muscle tissue which releases myoglobin, a protein that in high concentrations can damage the kidneys.

The severity of a burn is calculated by determining what percentage of the total Body Surface Area (BSA) is affected. This system, called the 'rule of nines', divides almost all the body into sections of 9 per cent or, for larger body parts, 18 per cent. The head and neck are 9 per cent, as are each arm including the hand, while each leg including the foot is 18 per cent. The front of the torso accounts for 18 per cent and so does the back and buttocks. Full-thickness burns that cover 2 per cent BSA are considered minor; those between 2 and 10 per cent are deemed moderate, while burns covering 10 per cent or more are considered major. After examination, Eric Brianchon was estimated to have burns over 70 per cent of his body and was the candidate for the first helicopter back to the mainland. As a rule of thumb, if the total BSA of burns added to the patient's age exceeds 100 then survival is unlikely.

In the hospital Dr Strachan supervised as patients were fitted with intravenous fluid drips and their burns carefully bandaged to stave off the serious threat of infection, always present with damaged skin. The atmosphere was hectic, with at times too many people anxious to assist and insufficient clear organization and delegation, but eventually things settled down into a semblance of order. It was then that Berriff stepped in and persuaded Dr Strachan to approach a patient and ask if he would mind answering a few questions on camera. Although Strachan was mindful of his duty of care to his patients, he was also aware of the magnitude of the event that had taken place and knew that a few men were keen to talk of their ordeal. He approached one man, who was unshaven and lying under a caramel-coloured blanket, and he agreed. At around 3 a.m., he gave the first account of the disaster.

> *Dr Strachan:* 'Could you tell me what happened to you? And what time if possible?'
>
> *Survivor:* 'Yes I woke up about half-past nine. I thought that was too early to get up so I just dropped back off to sleep again. Then at a quarter to ten or ten o'clock there was a loud bang and everything shook from the cupboards. Books came off the shelves and what-not. At first we thought one of the boats had hit the rig. Then when we got up we smelled the smoke. We just got dressed and went out to the corridor, and that was filling up with smoke at that time. We just started to look for a way out. I was in the accommodation I don't know for how long, just trying to find a way out. Eventually we got out onto the main pipe-deck area. Then we went round into this store and we were stuck in the store and there was another large explosion. So we were in there for I don't know how long. Then the whole pipe-deck level subsided where we were, it tilted to one side about thirty degrees. So it was a case of you had to get out of there. By that [time] the smoke cleared because the flames were not getting up past us and we could see the side of the rig, and we just crawled across there through the metal and just jumped over the side.'
>
> *Dr Strachan:* 'What height did you jump from?'
>
> *Survivor:* 'A hundred feet.'
>
> *Dr Strachan:* 'How long were you in the water before you were picked up?'
>
> *Survivor:* 'I was . . . I think about half an hour.'

> *Dr Strachan:* 'You have burns to your hands, how did you sustain that?'
>
> *Survivor:* 'That was crawling across the metal work towards the edge of the rig. Everything was just white-hot.'
>
> *Dr Strachan:* 'White-hot?'
>
> *Survivor:* 'The side was slippery because the soles of your boots were melting when you tried to walk across the metal. We just got to the edge – I was the first one across. I think there was about eight following me. A few of them went over first as they had life jackets on. I didn't have a life jacket on so I went after them and there was a few after me as far as I know. When I was in the water I swam to the boys with the life jackets and just hung on to them. And we just drifted out to some driftwood, climbed on board and stayed there until some rescue boat came along.'

The television camera was not universally welcome. While Berriff moved around the *Tharos*, spending time filming the men who were already moving the bodies arriving on deck, or getting shots of the care given to the wounded in the hospital, he was unaware of one man's contempt for his difficult but legitimate actions. Bob Ballantyne had been transferred from the *Lowland Cavalier* to the *Tharos* by the rope basket, an experience he described as follows:

> The comfort – the absolute humaneness of people who were not medical staff, just workers on it, one guy was either a scaffolder or a rigger. He was a tall chap, towering above me, and he called me son. He was probably half my age. He put his arm around me and he had a blanket. It was just that comfort. That human warmth. They took us down to the hospital. It was just chaotic. There were people being treated for burns. It was an emergency situation.

He then spotted Berriff with his camera, which angered him intensely: 'I got very upset. There were people with bloody video cameras and I thought: "people are dying – people have died and you bandits are going about with a video camera. That was rather nasty, I thought."'

Yet the anger he felt towards Berriff was balanced by the compassion he felt for the crew of the *Tharos*, who comforted the survivors and assisted them in intimate tasks, such as taking a man with two badly burned hands to the toilet. 'They would say: "Do you need the toilet? C'mon, I'll take you to the toilet," Ballantyne recalled. 'What an act of mercy – an act of concern.'

In the early hours the engines of *Silver Pit*, already troublesome, suffered a serious setback when the boat was rocked by an underwater explosion. Although no one was harmed, the lube oil pipe which fed the gearbox was fractured in the blast and then began to leak badly. This meant that Carson, the chief engineer, was called away from his live patients to tackle a mechanical one, although at first there was little more he could do to maintain motion than pour canister after canister of oil into the gearbox. When the oil eventually ran out, the captain found he had lost the pitch of the engine; this left him unable to manoeuvre the ship, which began to drift.

The vessel was impotent, but Ed Punchard was reluctant to rest. In the few hours before dawn the sky was at its darkest. From the deck he looked up to see the half-dozen helicopters flying in rigid search-patterns while scanning the sea with their fixed searchlights. Looking at the silver spear of the helicopter's beams, he realized that *Silver Pit* hadn't yet switched on their searchlight. A visit to the bridge resulted in an uncomfortable confession from the captain: although the vessel had a search-light, what it lacked were bulbs. Instead, all the captain could offer was a small hand-held signal lamp with an exceedingly short lead, which restricted the beam's panorama. The captain had only discovered these deficiencies after he had set sail and bore no responsibility, according to Lord Cullen's conclusions, for these failures. The captain and crew would later be praised for their bravery in maintaining their position and struggling with a crisis for which they were seriously undermanned.

Punchard set up the lamp and began to scan the waters, calling the chopper pilots if he spotted anything noteworthy. He was unaware, but out in the darkness floated the body of 'Lens'. It seemed Brian Lithgow had managed to escape from the platform and got into the water. However, his body would not be discovered until June the following year.

At one point Punchard and the divers attempted to launch the DOTI craft, a small emergency inflatable vessel, mandatory under Department of Trade and Industry rules, but their request was overruled by the captain on the grounds of the craft's condition. The craft's skin had bubbles the size of Easter eggs that would burst if brushed against ragged wreckage, resulting in more men in need of rescue. Stan MacLeod tried to get it started anyway, but the engine wouldn't work.

It took a couple of hours for Carson to get the engines started. The ship's drift had Gareth Parry-Davies worried that they were heading back towards the platform and he decided he would dive off if necessary, a dip he was spared by the direction of the current.

•

Dawn arrived at 3.30 a.m. The sun rising from the sea, taken as a universal symbol of hope, served also to illuminate all that had been lost. Standing on the deck of the *Silver Pit*, Ed Punchard felt the soft touch of the dead. He felt the comfort of his father's presence. Although lost ten years earlier, he felt that his father had been with him through the long hours of night, keeping his son safe, and was now departing, dissolving with the daylight. As the sky lightened with the return of a brighter light than Piper, the full panoply of the rescue operation was revealed. A flotilla of almost fifty vessels had converged, while the sky was scattered with helicopters, bringing their own problems.

By 04.00 a.m. Michael Clegg, the captain of the *Lowland Cavalier*, was swamped with radio traffic and could no longer cope with the communications. *Tharos* took control and Ashby, the deputy OIM, set up a search-and-rescue pattern for the various vessels now congregating on the scene. His guide was the merchant shipping search-and-rescue booklet, but the Coastguard recommended setting up two specific search areas: a 'distant' one, for those who might have escaped into lifeboats and drifted, and a 'personnel' one, for the area closest to the platform. The Coastguard provided the latitude and longitude for each area. Ashby then allocated each ship with a specific search area and instructed them to do parallel sweeps bearing north and south on the lifeboat search and west on the personnel search.

The *Silver Pit*'s engines were eventually fixed by 04.00 a.m., and it was first suggested by *Tharos* that the *Silver Pit* begin to sail back to Aberdeen, but, instead, it was decided to transfer the survivors into FRCs which would then take them onto supply vessels, whose greater manoeuvrability would make access to *Tharos*, and a helicopter home, easier. The first FRC contained two doctors who examined the injured, while praising Carson for his care.

In a bunk, tucked under a silver foil blanket, Mike Jennings had lain in a state of euphoria. The tangible thrill of being alive became like a form of delirium. In between the coughs and splutters he would smile. The coughs, however, grew sufficiently serious for Carson to suspect that he was suffering from secondary drowning, a result of swallowing too much water and smoke. The doctors agreed and so a helicopter transfer to *Tharos* was swiftly arranged.

Punchard was the last survivor to leave the *Silver Pit*. He spent a few minutes looking out at the boats moving up and down in strict search patterns and recognized two from previous jobs, the *Iolair*, BP's emergency support vessel, as well as the *British Enterprise III*, on which a close friend, Mike Toy, worked as a photo-technician. Before leaving he returned the work boots he had borrowed, exchanging them for a pair of flip-flops, and was able to shake hands with James McNeill and his crew when they were eventually brought back to the ship. Punchard then took an FRC to a supply boat where warm blankets, tea and buns awaited him. A few minutes later, he was transferred by basket to *Tharos*.

There each survivor had his own escort equipped with a woollen blanket. Punchard recognized his as Martin Richmond, another former colleague, who told him that he was supposed to have been under Piper Alpha last night, piloting a flying bell, but thankfully had been delayed.

After visiting the hospital for a check-up, Punchard went on to the cinema where the survivors had gathered, before accompanying Stan MacLeod (who was drawn by promise of a cooked meal) to the galley. There, to his delight, he met Keith Cunningham, who told him of his long swim to *Tharos*. He asked if there was any news of Barry Barber, but was told there was nothing as yet.

In the radio room, Punchard found a computerized list of Piper Alpha's occupants. An asterisk was pinned to those who had survived, like a badge of honour. There was nothing next to Barry Barber's name, nor those of so many others.

The exhilaration of the evening's events had transformed Punchard; his proximity to death had heightened his love of life to a degree which only therapy and an understanding of post-traumatic stress would later allow him to comprehend. On the helicopter ride home he felt invincible; he was enthralled by the bright blue of the sea and the greenery of the countryside. It was as if, after a life spent in black and white, he had been reborn into a Technicolor world. When he touched down at Aberdeen Airport he emerged with his chest puffed out, his hand on his hip and a broad smile on his face. This was a moment captured by a press photographer and although the picture was never published, when Punchard (after his subsequent slump) discovered it a few months later in the archives of the *Aberdeen Press & Journal*, he was so ashamed of his appearance that he stole the picture.

•

For Bob Ballantyne the experience was quite different. He was escorted with a group of other men on to either the sixth or seventh helicopter. The sky was light and after take-off the helicopter circled what little remained of the platform. Ballantyne looked out of the window and saw the leg to which, a few hours before, he had clung. He could not believe that there was nothing else left, that the platform where he had worked and the accommodation blocks where he had slept and all the friends he had known were gone, all gone. Then as the view was replaced by the blue of the sea, he rested his head against the window and began to sob.

PART THREE

ONSHORE

14. LONG NIGHT'S JOURNEY INTO DAY

The words on the page of the novel with which Molly Pearston had retired early to bed began to blur as her eyelids grew heavy with sleep. It had been a long day, and she was tired. A nurse at Tornadee Hospital, she had been on the early shift starting at 6.45 a.m., but tomorrow held the promise of a day off. After finishing at 3.30 p.m., she came home to a house which was quieter than usual, for Robert was offshore, as was Molly's eldest son, Lawson, and David, her youngest son, was staying with friends in Aberdeen. It was only herself and her husband for supper and afterwards, while he liked to stay up and watch television, she preferred to exchange the living room for the warmth of bed and a good book. Sleep now beckoned.

Molly turned over, put the book down on the bedside table, switched off the light and lay in the darkness. She had expected to drift off quickly, but instead she was gripped by a sense of foreboding and rising panic. Her heart began to pound and she became short of breath. In an attempt to shake off the sensation she sat up and tried to calm down. Looking at the clock, she saw that the time was just after 10 p.m. The feeling began to recede as quickly as it had washed over her, and she then settled down to sleep once again.

In the centre of Aberdeen, in a smart avenue of fine

homes, Kate Graham was also considering the benefits of retiring to bed with a book. She was about to enter her final week as head of public relations for Occidental, a job she once adored but had now wearied of, worn down by management's new focus on accountancy figures and politics over actual operations. In readiness for her last few days, she had spent the past hour or so preparing her final expenses, struggling to find the receipts. As she went through the paperwork at the dining-room table, Magnum PI played on the television in the living room next door.

She looked up to see the credits roll and then heard the pitter-patter of water coming from the recess. Concerned about a leak, she called her neighbours upstairs, who, upon investigating their loft, were greeted by a tide of water and phoned back with apologies and the promise of a swift repair. 'I've had enough,' she thought, and prepared to head for bed, but she was stopped by the phone. She picked it up thinking it might be Jim, her neighbour, but instead she was greeted by the voice of David Steele, the late-night reporter with the *Aberdeen Press & Journal*, the area's leading broadsheet.

He was straight to the point. 'Has anything happened off-shore tonight?'

'I haven't had word of it,' replied Kate, who said she would check.

She hung up, then dialled the number of Occidental's office and asked: 'Nothing's happened offshore tonight, has it?'

The first person she spoke to said: 'No – I haven't heard.' Then, in the background she could hear another voice say: 'an instigating call out.'

'What has happened?' she asked.

'There is word of an explosion on Piper,' came the reply.

'I'll be right there.'

Kate then called Steele back and told him: 'Something has happened. I know nothing. I'm leaving right now.'

When Steele asked if they should hold the front page, she said she had no idea and hung up.

She then rushed out, climbed into her car and began driving towards Occidental's offices. As she manoeuvred through the light evening traffic, she kept thinking: 'Don't let it be bad ... Don't let it be bad.'

•

News of the disaster had been broken to the Occidental Communications Officer at 10.03 p.m. by the *Tharos* and, in line with the company's emergency procedures, he initiated the cascade call-out system whereby senior personnel were called into the headquarters to man the Emergency Control Centre. The immediate calls went to Jim MacAllan, the Production and Pipeline Manager, who was the first to arrive at 10.21 p.m., followed shortly after by J. Coffee, Vice President of Operations, who was the official Onshore Emergency Controller. Yet as Coffee had been appointed to take over responsibility for the North Sea only recently, he had to rely on the advice of his senior colleagues in coordinating the company's response.

MacAllan and Coffee were joined in the Emergency Control Centre, situated on the third floor, by the senior managers of Production and Pipeline, Transport, Marine Operations, Loss Prevention and Drilling.

The room, which was set up by security guards, had two rows of five tables, each designed for two people and equipped with one telephone each. A number of whiteboards hung on the surrounding walls, on which new information could be written up as it arrived in. Each department had its own desk. While these included legal, admin support, human resources and health and safety, during the first few hours attention was focused on logistics, who organized helicopters and boats; engineering, who attempted to develop contingency plans; and Site Support, who in rehearsals had dealt directly with the OIM, but this night would communicate with Letty on the *Tharos*. The room had a single computer, invariably used for typing up press releases.

While the principal calls to Occidental staff were carried out by the Communications Officer, a lesser list of call-outs was the responsibility of security guards at the security lodge. Among their list was Grampian Police headquarters, who were notified at 10.08 p.m. that 'a major emergency has occurred offshore'. Under company rules, this was all the security guards were permitted to say. However, the police wished to know the identity of the installation, the nature of the incident, what action had been taken by the OIM and how many casualties were expected. The police officer then phoned back one minute later to authenticate the call and check for further information, but none was available, although the guard said he would call back with an update.

There then followed a period of confusion. While the Control Room at Grampian Police began to receive calls from the media seeking information on reports of an incident offshore, between 10.20 p.m. and 10.40 p.m. they were unable to contact Occidental as the line was constantly engaged. When the Police called the Coastguard at 10.25 p.m., the person who answered said he would phone back in a minute, which he did not. A second call to the Coastguard at 10.40 p.m. was barely more successful, with the respondent saying he was unaware of a fire on the Piper, but that there were reports of an explosion and that a Nimrod and six helicopters had been scrambled.

Chief Inspector Ian Gordon had just arrived home from a successful evening's game of bowls when the phone rang at 10.43 p.m.
The ringing of telephones would become his abiding memory
of a disaster that would consume the next year of his life, and
lead him to associate a phone's incessant ring with nails on a
chalkboard. The chief inspector had been appointed Grampian
Police's Oil Industry Liaison Officer in November, prior to which
he had been head of Forensics. The new role was a full-time
position that involved working with each oil company on issues
of safety and security as well as coordinating the regular NORAX
– North Sea Offshore Exercises – sponsored by the Department
of Energy and involving the coastguard, police, air force, navy
and hospital services. Previous exercises to date had involved
accidents on a platform that resulted in only a few 'deaths'.

When Inspector Gordon was told of the confusion, and the
previous attempt to reach him at 10.25 p.m., he immediately
ordered that a senior officer be sent to Occidental's offices to act
as liaison, a reciprocal agreement which was part of existing
contingency plans. A second police officer was to be sent to the
Coastguard headquarters to act as the force's eyes and ears.
Gordon then drove straight to police headquarters, which he
reached at 10.55 p.m.

Kate Graham arrived at her office and got out the emergency
file which contained a prepared statement – already carefully
worded – on which it was only necessary to add the name of the
rig or platform. As she ran down the corridor to another room,
she passed a colleague and asked: 'What is it?' When he said it
was a rig, her initial reaction was: 'Good, if it's a rig it won't
have as many men as a platform.' The phones were already
ringing as she arrived, as the first news report of an offshore
incident had been carried at the tail-end of ITN's *News At Ten*.

A few minutes later she learned that it wasn't a rig but a platform: Piper Alpha.

At the BBC's Aberdeen office on Beechgrove Terrace, next to the verdant plot from which Scotland's most popular gardening programme was broadcast each week, Jane Franchi was hitting the phones hard. The BBC's Northern correspondent had been called in when the first reports began to swirl, and initially she thought it would be a minor incident. Her expectations were low, as it was the third report of an offshore explosion that week, but her first call to the Coastguard – the media at least were having no trouble getting through – confirmed that this was being treated as a 'major incident'. Together with Eric Crockhart, a reporter with Radio Scotland, they divided up the calls and began to scour for as much information as possible. Shortly before 11 p.m., she was told that Aberdeen Airport was reopening; at that time the airport, like a strict landlady, closed its doors and rolled up its runways promptly each night at 9.30 p.m., forcing any flights which were delayed to reroute to Edinburgh, to the disgruntlement of passengers.

Franchi received an insight into the true reality of the situation 110 miles out to sea when the Coastguard – shortly after receiving the latest situation report from Rescue 01 – told her that Piper Alpha was 'burning from the top of the derrick to sea level'.

This was new information and Franchi was adamant that it had to get on the air. As the evening news had already passed, the only available outlet was *Newsnight*, BBC2's flagship current affairs programme, which on this evening was devoted to the judicial inquiry into the Cleveland child-abuse scandal, when nearly 100 sets of parents were wrongfully accused of sexually abusing their children based on the controversial examinations made by the paediatrician Marietta Higgs.

Franchi managed to convince the programme's producer of the urgency and her report – the last item, spoken over the telephone and less than one minute long – brought the horror into the nation's living rooms and would light the touch paper on the public's response. At the time, 25,000 men worked offshore, but many of their wives and girlfriends were unaware of the name of their actual platform.

Dr Graham Page, head of Accident and Emergency at Aberdeen Royal Infirmary, had watched with concern as the first report came through on ITN on the portable television wedged among the clutter of files, books and videotapes that threatened to overwhelm his tiny office. In the case of any emergency offshore it was common practice for the hospital to be informed by police or coastguard, but they hadn't heard anything. Nonetheless, it was decided to implement the infirmary's major accident plan. The first set of calls brought in extra medical, nursing, administrative and portering staff, as well as those required to staff radiography, records and the telephone exchange.

Given the potential population of a large oil platform, infirmary staff anticipated dealing with as many as 200 men, and while no one expected that number to require hospitalization, they succeeded in freeing up 192 beds. Accident and Emergency, the intensive-care unit and the plastic surgery ward were cleared of all but the most serious patients, while a burns dressing and minor procedures area, which was not due to be taken over from outside contractors for another five days, was rapidly swept into use. An additional documentation and record-retrieval point was set up, extra porters were put in place and fresh blood supplies called up from the Blood Transfusion Service.

The hospital would also be provided with extra assistance. Professor Sharpe, who had conducted a number of complicated skin grafts on burn victims in the aftermath of the Bradford fire, had earlier in the evening completed a lecture on his procedures in Leicester, and was in a city centre bar when he spotted a television report on the unfolding disaster. He made an instant decision to offer his and his team's assistance, and within a couple of hours the RAF had arranged to collect them at East Midlands airport and fly them north by helicopter.

Shortly after 1.30 a.m. the plans were complete; the hospital was now in a state of heightened readiness; but just as in battle, where no strategy survives first contact with the enemy, the Royal Infirmary – while prepared for a high volume of patients – would be surprised by the number of relatives who even now were hurriedly climbing into cars from across not just Aberdeen and the north-east, but Scotland as well as parts of England.

Grampian Television, the local ITV network, decided to broadcast brief news reports every hour throughout the night, and reported that survivors would be arriving by helicopter at the Royal Infirmary. Immediately the hospital's switchboard lit up as the operators explained again and again to each caller that as yet there were no names.

The lights that ringed the hospital's concrete helideck, illuminating the yellow 'H', were switched on. When relatives arrived, they were shown into a small room in the A&E department, which was normally used for teaching medical students. The first to arrive were locals from Aberdeen and the surrounding towns such as Peterhead. Wives and girlfriends had dressed quickly, leaving children with neighbours and family in order to make the journey. Alan Reid, the hospital's head of public relations, believed it was best to try keeping them together and away from the press, who had also begun to gather outside. When the number of relatives rose beyond the room's capacity to cope, they were moved to the chapel, where the neat rows of chairs were divided up as the families formed small groups for mutual support. So little information was available – it would be hours before the hospital was provided with a 'persons-on-board' list – that all everybody could do was sit and wait, and sip from the cups of tea and coffee provided by the catering staff.

At the heart of the room was the Reverend Alan Swinton, a Church of Scotland minister of quiet charisma and tangible faith, who was the infirmary's principal chaplain and among the first to arrive. He moved from group to group, asking people's names and listening to the stories of where they had come from, of what their husband, son or boyfriend did offshore and of their rising fears that all too frequently ended in tears.

When Alan Reid returned to the chapel to check if there was anything else he could do, he was asked about access to a television, so he crept into a neighbouring ward where the patients were asleep, unplugged the TV set and wheeled it off.

•

At Occidental at midnight, Kate Graham was struggling to get clear up-to-date information on the scene. As there was only one satellite line linking the office to the *Tharos*, she did not expect to gain access to it for means of updating the media. However, she did expect her colleagues in the Emergency Control Centre to update her whenever possible. So when she visited the Centre and saw details written up on the board about the different aircraft dispatched to the scene, she lost her temper and shouted in no uncertain terms that she needed the facts. A few local journalists, whom she knew and trusted, had arrived at the front entrance and so she let them into the office canteen, secure in the knowledge that they would not go wandering off. They later told her that she looked as pale as a sheet.

In the early part of the evening she and her team were brooding on the possibility that two or three men might have lost their lives. It took time for the totality of what had occurred to penetrate through the office and its staff. Jane Stirling, one of three structural engineers called out to assist with contingency plans, spent the night fetching detailed drawings of the rig, but then was told not to bother: 'Module D has gone.' She responded: 'Gone? What do you mean *gone*?' In the space of a few hours, the schematics had become historical documents referring to a platform that no longer existed.

At one point, in the early hours, a man walked into Kate Graham's office and stood in front of the whiteboard. When he left, she could see that he had written the figure '190' followed by the word 'missing'. Tears began to pool in her eyes as the emotion of the moment rose in her throat, yet she knew that if she were to break down it would trigger the collapse of her team, already emotionally drained by gruelling encounters with angry journalists demanding information her staff could not yet provide. She thought: 'I can't.' Instead she turned round, substituted a hope that was sincere if ultimately futile and said: 'There are bound to be some boats around with the men on board. It won't be as many as that.'

On the sixth floor of police headquarters a major incident room had been set up, filled with the incessant trill of the telephone. Two casualty documentation teams had been called out by Inspector Gordon, one to staff the major incident room while the second attended the Royal Infirmary. The scale of the incident was revealed at 11.35 p.m. when the police officer acting as liaison with the Coastguard called to say that the entire installation was ablaze. An hour later the police officer at Occidental reported the platform's collapse.

While Occidental finally provided the police with the 'pob' (persons-on-board) list at 12.45 a.m., it was inaccurate and included the names of seven men who had transferred to *Tharos* in the hours before. It was also difficult to read as it was arranged alphabetically under the heading of each of the 28 different subcontracted companies, thus requiring staff to read through each company rather than being able to go straight to the letter of the surname about which a particular member of the public was enquiring.

Gordon first thought about what to do with the dead at around 2 a.m. In the previous hour they had learned that only 41 survivors had been recovered – though the number would rise – and a single body. He knew they were looking at the recovery of as many as 160 more. The mortuary at police HQ was equipped to hold 12–15 bodies, and while there was a dedicated mortuary for emergencies he believed this was inappropriately located in an old council yard amid a housing estate in Northfield. He decided to raise his concerns with the Chief Constable, Alastair Lynn. What if the press got access to the neighbouring flats and were able to photograph the arrival and departure of bodies?

The solution was drawn from Gordon's previous experience as head of Forensics. In November 1986 a Chinook helicopter returning from the Brent Oil Field to Sumburgh in the Shetland Islands had crashed almost three miles offshore, killing 45 of the 47 people on board. On recovery the bodies had been housed in a hangar at the island's airport which they turned into a temporary mortuary. Lynn having agreed that they should try to set up a similar arrangement at Aberdeen airport, the local superintendent was dispatched to talk to the airport authorities, who immediately agreed to empty a hangar currently storing snow-clearing equipment. Gordon was also keen to draft in Kenyon Emergency Services, a specialist team of morticians with whom he had worked on the Chinook disaster. Aware that the company would wish to know who would be paying their fee, he had a quiet word with the liaison man whom Occidental had sent to the Control Room. Unfortunately Occidental's representative was insufficiently senior to make an immediate decision, but after conferring with senior management he returned to say that whatever the police required Occidental would fund.

Gordon had the home telephone number for Peter Kenyon, the company's founder, and woke him at around 3 a.m. Kenyon agreed that the hangar was an excellent idea and that he would begin rounding up a team immediately.

Sitting at his desk in the Control Room, which measured 10 square metres and was staffed by some officers answering phones while others were inputting each new piece of information into C.A.P.S. (the Computer Assisted Policing System), which had recently been introduced, Gordon knew it was necessary to gear up for a major investigation that, if foul play was detected, could lead to a charge of mass murder. Under government legislation it was the police's role to take the lead in casualty documentation, the identification of bodies and the management of an emergency mortuary, as well as gathering accurate statements from each of potentially hundreds of witnesses.

The next call was to Detective Superintendent Alastair Ritchie, the deputy officer in charge of CID. He was awoken at 3.15 a.m. and asked to supply a team of officers to collect statements from the survivors, who were scheduled to arrive at the Skean Dhu Hotel at Aberdeen airport once they had been cleared by Infirmary staff. A second team was to prepare to fly offshore as soon as possible to take official charge of body recovery, while a third team would run the mortuary. Ritchie knew that he was running a criminal investigation, and that each witness statement or link in the chain of evidence could be scrutinized by a court of law.

After Ritchie had been briefed, Gordon was informed that the first helicopter carrying survivors was expected to touch down at the Royal Infirmary at any moment.

•

In the soft twilight that precedes a midsummer dawn the sky was a pale blue, dotted with a few faint clouds that over time slowly teased apart. For the journalists who had gathered on the grass beside the Royal Infirmary's helideck, there was little to do but watch the sky. In an attempt to keep the crowd controlled, Alan Reid had taken to appearing outside every fifteen minutes to offer even the most minor update about the hospital's preparations. The announcement that the first helicopter bearing survivors was expected to touch down at 3.30 a.m. was more substantial, and had the press pack anxious with anticipation. Inside A&E, Reid had wiped down a whiteboard and wrote up the helicopter's estimated arrival time – the first, he hoped, of dozens.

In the past few hours the number of relatives gathered in the chapel had risen to almost two hundred; each new face was cornered as they entered by earlier arrivals who were anxious to check if there was any more news. Hard facts were still scarce. While the police and Occidental now knew the full extent of the catastrophe, this had yet to leak out. Meanwhile, each person's hopes poured into this vacuum of information. Reid and the Reverend Swinton were anxious that the helicopter's arrival should not trigger a stampede amongst the relatives, whom they placated with assurances that as soon as the men's names were known to them they would return and inform those who were waiting. The chapel's windows overlooked the helipad and as the thrum of the rotor blades overhead was heard and the dust blew in the air, kicked up by the down-draught, women rushed to press their faces against the glass to see if their relative was coming home.

As the helicopter touched down an ambulance reversed into position, its back door open to receive the first patient, Eric Brianchon, swaddled in blankets and with an oxygen mask over his face. Brianchon was carried on a stretcher and loaded into the ambulance, which set off on a journey of 200 metres to the A&E's front door, with the siren on.

Over the next three hours eleven helicopter flights touched down, carrying 63 people, 21 of whom were admitted to either the plastic surgery and burns unit or the A&E ward. As the men were disembarking from one flight, a woman at the chapel window began to shout that she could see her husband. The Reverend Swinton tried to persuade her to wait, saying that he would check first and let her know, but she insisted on running down to the A&E, only to discover that she had been mistaken. She then crumpled down on the floor in tears.

The Police Casualty Documentation team at the hospital were providing hospital staff with information regarding the helicopter arrivals, and recording the details of those who were on each flight. Police officers were also working closely with the hospital chaplain with regard to the information he was giving out to relatives.

•

Molly Pearston was woken up by the telephone. She had not heard her husband come to bed and knew that, as a heavy sleeper, he was unlikely to stir and answer it, so she went downstairs, lifted the receiver from the hall table and then sat on the bottom stair.

A voice said: 'Hello – is that you, Molly?'

It was David Adams, a mechanic who had trained with her son, Robert, at Thomsons of Cult where he had learned his trade.

'Is Bob working on the Piper?'

Molly looked across at the clock. It was 1.15 a.m.

'Oh, Piper, yes. I think so, why?'

He said not to be alarmed, but there had been an incident on

Piper, a fire. He thought it might not be too bad, and that Grampian Television were carrying reports.

After she hung up, Molly ran upstairs to wake her husband and tell him that she was going to start ringing her son's boss. Mr Pearston, barely awake, nodded in agreement and then fell asleep again. Over the next couple of hours Molly repeatedly tried to get through to Grampian Police, as well as to her son's boss. She even tried to phone Grampian Television, seeking further information. At around 4 a.m. she saw on a TV news report that the men were being brought to the Royal Infirmary. She called her youngest son, David, at his friend's house, and he agreed to go straight to the hospital. She then waited until after 6 a.m., before calling Robert's girlfriend, Lesley, who lived at home with her parents, thinking she would need her sleep for what was to come. When told, Lesley was upset and insisted on meeting David at the hospital.

The atmosphere they encountered there was a mixture of anger and stunned disbelief. Occidental had failed to send any-one down, perhaps in fear for their personal safety since one woman, frustrated by the lack of information, broke down in the A&E and screamed: 'Those bastards have killed my husband . . . Those bastards have killed my husband!'

During the long wait for the first helicopter Jane Franchi began chatting to the brother of a survivor, who promised to come back and give her an update once he had spoken to him. When he returned he said that his brother was not making any sense: 'He says the platform has blown up – all of it – that it is all at the bottom of the sea. That can't have happened.' He said he thought his brother must be delirious.

When Bob Ballantyne climbed out of the helicopter he was immediately wrapped in a blanket and escorted to an ambulance, but as he entered the hospital the families pressed around him and began asking if he knew what had happened to different men. He couldn't bear to tell them what he had seen:

> I knew that people were desperate for information and the only way that I could cope with it – it might have been a coward's way out – but I just said, 'There are other helicopters coming.' They were asking for names. Did you see Jimmy or John? The only thing I could say to them was, 'There are other helicopters behind us.' I just wanted to get away from them. I know it was lies I told, but it was the only way I could cope.

Shortly after 7 o'clock that morning, Alan Reid looked at the whiteboard and saw that it had not been updated with the estimated arrival time of the latest helicopter. Furious, he demanded to know when it was due. He couldn't quite believe it when told that there would be no more flights. (It would be weeks before the board, detailing each of the 11 flights and their occupants, was wiped clean.) This news was broken to the relatives in the chapel by the Reverend Swinton:

> 'There must have been more than 100 people gathered at that time. There was just a numbed silence when I told them. I thought then that everybody would go home, but very few did. They didn't want to go away from here. I think they were finding comfort in being with each other, and to some extent some felt there was a real world out there, whereas they were protected here in the hospital.

The protective walls of the chapel could not hold back the reality for long, however. No more helicopters meant no more hope, and numbed silence gave way to quiet sobs and a few angry scenes as some women raged at hospital staff before collapsing into their arms with exhaustion and grief. Those doctors and nurses not immediately involved with treating the most seriously injured then had time to take stock of what had happened. When he had a moment, Dr Page sat quietly and stared into space.

No one knows how the story of the ship of survivors started; it was born in hope and nurtured by each who heard it. Shortly after they arrived Lesley and David were told that a trawler, perhaps Russian, had picked up between 80–90 survivors and was sailing for Peterhead or Fraserburgh harbour, but that the vessel's radio was broken. In the hope that Robert might be on-board, Mr Pearston called friends who lived in both towns and they agreed to go to the harbour and look to the horizon for a ghost ship that would never come to port.

•

It was rare for Pat Slater to go to bed without watching the late-night news, just in case there had been an accident offshore, but the latest record from the Open University had arrived as part of Bob's course and she wanted to record it onto a cassette. A music graduate from Aberdeen University, she listened to Mozart's *The Marriage of Figaro* and wrote out a series of accompanying notes pointing out the changes in the musical passages. She then went to bed and never heard the phone ringing in the living room. Bob had asked two nurses to try to reach her, and when they failed for a third time he decided to joke with them, saying 'She normally goes to the dancing with her sister on a Wednesday night. She's probably not home yet.'

The nurses didn't get the joke and looked appalled.

Pat was awoken at 6.30 a.m. by two police officers ringing the entry phone on their third-floor flat and asking if a Robert Ballantyne lived here. Her first reaction was: 'What has he done now?' When they came up the stairs they said that there had been an accident, but Robert was one of the lucky ones. She waited until after 7 o'clock to call her sister Anne, who had a one-year-old son, David, but who came straight round and accompanied her to the hospital. Anne was a psychiatric nurse and as they drove she told Pat that Bob would need to talk about what had happened. She advised: 'Get him talking and when he's finished, get him talking some more.'

They found the hospital in chaos, with no one able to tell them where Bob was; first they were directed up to the chapel, where they walked into a prayer service going on for those whose men had not come home. 'It was very inappropriate for us to be there and I just wanted to find Bob,' said Pat.

Eventually the sisters were shown into a small room where they found Bob, still in his orange overalls, his face smeared with Vaseline and his hair 'greased up like a gangster' from the oil spill. Feelings of agitation, anger and annoyance had rolled through him as he waited, but when Pat and Anne came through the door, the only thing he could think to say was to reduce the catastrophic to the comic. Pat had bought him a new pair of trainers for their holiday. Now he said: 'I'm terribly sorry sweetheart. I've lost one of my trainers!'

•

Once the relatives had begun to depart, Reverend Swinton was called to the bedside of one survivor who was badly burned and scheduled to undergo the first of multiple skin-grafts. He pulled up a chair and began to pray for him, for his family and for the doctors into whose hands he would soon pass. When Swinton had finished he said: 'Amen.'

Then the man's cracked lips began to move. Swinton leaned forward and, as he recognized the words, he could feel the emotion of the evening, pressed down by professionalism, begin to find release. It was as if the words encapsulated all human frailty and acted as a brief balm on the wounds of this night. On the *Tharos* the bodies were being brought on-board under the watch of officers from the CID who attached a serial number to the dark plastic bags in which the men were placed. The ships continued to patrol strict lines of latitude and longitude in a search that would yield not the living but only the dead. In the sky helicopters flew by. And still Piper Alpha burned, quieter now than before, but enough to produce a smoke trail that rose to 40,000 feet.

In the hospital Reverend Swinton listened as the words of this man who had passed through an inferno rose up and blanketed all that had come to pass, penetrating the deep waters and reaching down to touch those who floated in the dark. He began:

> Our Father, Who Art In Heaven,
> Hallowed be Thy name,
> Thy Kingdom Come, Thy Will be done,
> On Earth as it is in Heaven . . .

Swinton then joined in: 'Give us this day our daily bread . . .'

15. THE PRIME MINISTER,
THE PRINCE AND THE 'HAMMER'

Dr Armand Hammer ordered that his gold Rolls Royce and usual London driver be waiting at Heathrow Airport to greet his private jet. There, accompanying both, was Ranulph Fiennes, the explorer, who – when not circumnavigating the globe in a novel manner – acted as executive consultant to the chairman of Occidental Petroleum. The pair had been introduced almost a decade before by Prince Charles, who, hearing that Fiennes was struggling to secure the necessary fuel to undertake his Transglobe Expedition (which involved journeying around the world on its polar axis using only surface transport, a journey of 52,000 miles) had recommended that he approach Dr Hammer. The wily executive would go on to persuade a fellow tycoon to fill the team's tanks, but did recognize Fiennes as a kindred spirit and later offered him a position as his personal PR man in Britain. The figure who came down the steps from his jet late on Thursday evening and climbed into the leather-upholstered back seat of the Rolls was straight-faced. Fiennes never knew Dr Hammer to betray his emotions, but even he was surprised by his control under the circumstances. During the eleven-hour drive to Aberdeen, Dr Hammer rarely spoke, preferring instead to look out of the window or nap.

The world had woken to the consequence of the worst offshore oil disaster in history. In every country news bulletins were broadcasting footage of the smoking ruin of Piper Alpha. Peter Morrison, the Energy Minister, had already accompanied John Brading, Occidental's UK president, to the *Tharos* from whose deck they surveyed the devastation. Meanwhile, Cecil Parkinson, the Energy Secretary, had announced that two major public inquiries into the disaster would be carried out. The Department of Energy and the Health and Safety Commission would investigate what had gone wrong, while a wider public inquiry (to be chaired by an eminent legal figure) would follow. However, there were more immediate concerns. Dr Hammer had called Red Adair at his home in Houston and the veteran firefighter had agreed to put together a team and fly out as soon as possible to begin work on putting out the wells that continued to burn. Dr Hammer had believed it was necessary to visit Aberdeen as quickly as possible, as he told a press conference on Friday afternoon: 'I am the chief executive officer of the company and the buck stops at my desk.' Yet he was not alone in his plans to visit the injured.

For the staff of the Royal Infirmary, Friday 8 July was to be the day of the Hammer, the Prime Minister and the Prince and Princess of Wales. Dr Hammer was the first to arrive, the gold Rolls Royce pulling up outside the hospital shortly before 1 p.m. While Sir Ranulph Fiennes used a mobile phone the size of a house-brick, trying to arrange with St James's Palace for Dr Hammer and the Prince and Princess to meet, his employer donned a set of surgical scrubs to speak with the survivors. Erland Grieve lay with his left arm raised in a foam sling, the right arm thickly bandaged and right hand sealed in a plastic bag. Dr Hammer looked down at him, his eyes hooded behind thick, dark spectacles, but his face expressed distress.

Outside, a thicket of 300 journalists had gathered from

around the world. The press in Norway, a nation buoyed by oil exploration, had chartered an aircraft, while camera crews had flown in from Japan and North America. After the departure of Dr Hammer, they were not required to wait long before the arrival of Margaret Thatcher, the Prime Minster, accompanied by her husband, Dennis, and Malcolm Rifkind, the Secretary of State for Scotland. Mrs Thatcher spent an hour touring the hospital, after which she emerged and went to the hospital's helideck, where those pilots involved in ferrying the injured had gathered. Dressed in black, and wearing a small silver brooch, the Prime Minister said:

> These are people with great inner strength and great capacity to endure a disaster. I was amazed at how high their morale is. They realize the terrible ordeal they have been through. They are talking about their experiences and the doctors are encouraging them to do so ... This whole explosion is clearly very deeply shocking in its magnitude and the numbers of lives lost.

She then announced the creation of a disaster fund to which the government would contribute one million pounds, a sum matched by Dr Hammer, who she went on to describe as: 'a marvellous person whom we have known for years.' She went on to say: 'He is a man who has by his own efforts done well. He makes certain that as he himself has prospered so he helps others, and I am sure he will look after the families.'

The Prince and Princess of Wales, whose Royal Flight north had brought up a new machine for perforating holes in skin grafts to replace one broken at the Royal Infirmary during the previous day, were the last to arrive at the Royal Infirmary and, as Alan Reid noted, did so in an ordinary Rover. The couple immediately passed a letter to the Lord Provost of Aberdeen, which read:

My wife and I were deeply shocked by the terrible tragedy which occurred in the North Sea and we wanted to convey our most heartfelt sympathy to the relatives and friends of those who so tragically lost their lives. Our thoughts and prayers are with them at such a difficult time.

After speaking to the survivors, the couple paid a surprise visit to their rescuers.

The *Silver Pit* had just sailed into Aberdeen harbour and tied up when a police squad climbed on-board accompanied by a sniffer dog. James McNeill asked what they were looking for and when told bombs, he replied: 'Bombs! Bombs! We've just left a fucking great bomb!' The atmosphere relaxed when told about their royal visitors, prompting each man to rush off and deodorize with liberal quantities of Old Spice, which led Princess Diana to say: 'Someone smells good.' When asked about the conditions in which they had found themselves, one man replied: 'If Hell is that hot, I'm going to be good!'

•

In order to cope with the large numbers of journalists, photographers and camera crews, the press conference was held in a gymnasium hall close to Occidental's headquarters. As Kate Graham escorted Dr Hammer into the room, the press surged forward and she felt the elderly chief executive slip a protective arm around her. To judge by his appearance, she felt she should have been the one to extend a comforting arm. In all their many meetings over the past twelve years, Armand Hammer had been warm but inscrutable, a titan confident of his authority and power. Today she saw the reality behind the role; he was an elderly man, shocked by the events of the past 48 hours and shaken by the sight of the bandaged men in the hospital wards. It was as if a central pillar supporting him had suddenly been knocked loose.

Yet when he sat down he steeled himself and attempted to project an image which was described by one reporter as 'stern and strong-voiced'. During close-ups the camera revealed that his eyes were moist. He began by stating:

> I thought it was my duty to come here. I am the chief executive officer of the company and the buck stops at my desk. This is the greatest tragedy that has ever happened in the oil industry in the world, and I thought it required my personal attention.

He went on to explain that the company had already appointed an investigative committee headed by Glenn Schurtz, a senior executive, and that it would 'leave no stone unturned to find out what happened. His investigation will be unbiased and he will do everything possible to find out what happened.' Dr Hammer stressed that the platform was subject to regular checks by the Department of Energy, and they were satisfied that safety procedures had been in order. However, he conceded that there appeared to have been a gas leak in C Module.

The questions then turned to the payment of compensation to the families of the dead and the injured, and while Kate Graham felt such questions were 'silly' at that stage, lawyers in Aberdeen were already discussing plans for joint action. The next day's edition of the *Press & Journal* carried an advert from the Law Society of Scotland, discouraging families from agreeing to 'hasty offers of settlement' and encouraging them to contact the society for help and guidance. Dr Hammer, in no mood to haggle in public over the cash value of a life, insisted that the company's own insurance policy was 'very liberal' and that they would work closely with all the contractors to ensure a fair settlement.

'We want to make sure that everybody is treated fairly,' he stated.

While dismissing the possibility of future litigation, he held himself ransom to fortune and the future findings of a public inquiry. 'I think we have taken every precaution we know how. I do not think there will be action.'

Shortly after the press conference, Dr Hammer climbed back into his Rolls Royce for the long journey south. He never returned to Aberdeen.

16. SIFTING THE ASHES

On arrival at Aberdeen Airport, Red Adair was briefed by Leon Daniels. The man who had set up Occidental's operations in the North Sea and was now the company's global drilling manager, based in California, had caught the first available flight back. He knew little more than had already been reported on the news and subsequently. Adair and his two-man team, Raymond Henry and Brian Krause, flew out to the *Tharos*. As they approached, Adair asked the pilot to move in close and slowly circle the platform; as he peered down at the smoking ruins, which now sloped at a 45-degree angle, he knew it would hold.

'It's not gonna fall, not if it's stood this long' he said to Henry and Krause. 'I can tell you that for sure.' He could also tell them that getting to the wells was going to be a 'bitch'.

Each of Piper Alpha's 36 wells was equipped with a downhole safety valve, a metal flap positioned inside the production pipe at a depth of 800 feet. The valves were connected, via a ¼-inch stainless-steel line, to a hydraulic pressure source which kept the flaps open. In the event of an accident that broke the line and cut off the pressure, the flaps would slam shut like the lid on a paint tin. There were fires on the platform, but it was not known how many – if any – of the flaps had shut. In order to find out, Adair and his team first had to figure out how to gain access to the wells.

Once on *Tharos*, Adair asked Letty to move the vessel to

within 60 feet of the platform. He then joined Henry and Krause in a steel basket connected to *Tharos*'s crane, which raised them up to the platform's flare boom. During this initial inspection Adair decided that the platform's three heat shields would have to be removed to allow the *Tharos*'s hydraulic gangway to be fixed in place. This would also enable the smoke still covering the platform to properly disperse. Then the hard work could begin. The well bay was buried in debris, and only after each piece had been removed would they know how many wells were still ablaze and required to be 'killed'. Looking over the heat shield and down into the well bay, Adair thought it resembled a jagged junkyard, slick with oil and crooked edges, and he began to wonder if he could keep his feet.

In the cabin set aside for use as the team's office, Adair explained that although he was confident that the platform would not collapse, they would be keeping meticulous records of what they uncovered, so that, in the event that he was wrong and the platform did fall, they could begin again with a team of divers. Each man knew the trouble it would cause: a subsea kill could take two to three years to complete. The first task was to get onto the platform, secure themselves with safety-ropes and then guide the crane in to clear the debris. Adair said he would speak to the machine shop about grinding them some grappling hooks. Work would start the next morning.

That evening, Red Adair decided that he would not set foot on Piper Alpha. Although he would direct the entire operation from the *Tharos* and watch over his men from the basket, he knew that the surface was too treacherous for a man of 73. To do so would be to act against the instincts that had kept him and his team alive during forty years and hundreds of fires. Yet it was not an easy decision. He felt a deep sense of sadness at what had occurred and, although he had predicted a disaster in the North Sea, he was disappointed that the *Tharos* had been unable to do more. The vessel was modelled on the *Phillips SS*, the first semi-submersible firefighting support craft, a 4ft model of which – built at a cost of $25,000 – sat in the lobby of Adair's warehouse.

As he pondered the decision he wandered around the *Tharos*, lingering on the upper deck where he could taste the smoke before heading down to the submarine room, which was part of the vessel's saturation diving system. The captain had decided it was too dangerous to allow divers to explore the wreckage for bodies, so instead the retrieval was carried out by remote-operated vehicles (ROVs). The ROVs were propelled by small thrusters controlled by a pilot using a joystick on the surface, who watched the undersea world on a television monitor. When the powerful lights fitted to the vehicles illuminated a hand or a face, the larger of the two ROVs, which was equipped with mechanical arms, was sent in to fetch him. CID officers had already spoken to the pilots and insisted that they maintain precise logs detailing who was at the controls and at what time a body was recovered.

After watching the pilots at work for a few minutes, Adair went back to bed.

Over breakfast the next day, he told his two-man team that they would be in charge of controlling the platform: 'I'll supervise the operation. I'll make sure the welders and machinists build those hooks just right for you. I'll make sure there's plenty of water on you all of the time. But I'm not gonna be over there on the platform with you. I'll be here on *Tharos*. We need a man here to make sure everything runs smooth.' Later Krause would conclude that this operation was Adair's most difficult as he watched from the basket while his men dodged flare-ups, for every few hours the wells blew out and burned furiously for 15–20 minutes before exhausting the available oxygen and dying down.

For four weeks, the team worked fifteen-hour days. Once the

heat shields were down, they began at the end of the platform free from fires and cleared away the debris, each piece of which was tagged by the police as evidence and shipped back to a storage yard outside Aberdeen. The machine shops on *Tharos* worked continuously on whatever the team required, including a grappling hook made out of 9⅝-inch pipe casing strong enough to pull 70 tons; A-frame supports on which they could stand; and a 15ft platform made out of 12-inch steel H-beams. They were unable to work for three days straight when the wind hit 75 knots and the waves reached 35 feet.

There were narrow escapes. A few minutes after the pair had returned to the *Tharos*, P1, the well on which they had just been working, blew out and caught fire, triggering the largest blaze since 6 July. The ignition of P1 and the length of time the team had already taken meant that Occidental were concerned that Adair might not be able to shut down the wells, therefore a back-up plan was put in place. A drill ship sailed out with the intention of drilling a relief well to P1, which would divert the oil and cut off its fuel supply. Adair agreed to the plan, but was adamant that his team could cap the wells before the tipping point was reached at which, economically, it made more sense to pull out the firefighters than the drill ship. Gene Grogan, Occidental's vice-president in charge of engineering, was quoted in *The Times*: 'We have not put a finite time scale on Mr Adair's operation. But it is not infinite either. It certainly does not look very good.'

The team had discovered that there were seven wells on fire, but that the safety valve was not the issue. Each well had a space called the annulus. This was a gap between the outer cylinder – the well-bore casing – and the inner cylinder, either the drill pipe or the production pipe. While the safety valve had closed over the inner pipe, it did not close the annular space and it was from this that the fires sprouted. The team's break came when the wind blew in just the right direction, allowing Adair to land a water jet right onto P1. After the flames were briefly extinguished Henry and Krause moved in and stuffed in a packer – a sealing device that expands inside the well-bore – after which they pumped fluid five miles down the inner cylinder to the bottom of the pipe, at which point it rose back up through the annular space, thus killing the well. Knocking out P1 made the remaining wells easier to manage, and within the next two weeks each one had been permanently sealed with cement.

During the period when Adair was at work offshore, his celebrity in Britain had grown to the point where Kate Graham was having to deal with requests from individuals who appeared to have forgotten the tragedy that lay behind his efforts. A BBC show asked if they could fly him to London, so that he could share his stories on the sofa, while parents wrote asking for signed autographs for their children. It was a relief when, after a final press conference, she escorted him to the airport for the flight back to Houston.

17. RAISING THE DEAD

The fires were extinguished on Piper Alpha, but they continued to burn in the minds of those who lived through that night and with the families of those who did not. The grey granite building of St Nicholas Kirk off Union Street in Aberdeen was the site of a memorial service that united the two groups. Ed Punchard had purchased a bundle of French magazines at Heathrow Airport and planned while back in the city to check on the progress of Eric Brianchon, who, according to recent reports had been rallying round. Sadly, he died the day before the service, the announcement coming too late to slip his name into the list of the dead on the printed order of service.

As every seat in the cathedral was filled, thousands of people gathered outside, clogged the streets and listened as the service broadcast on loudspeakers began with the words: 'We are here to remember those who died on the Piper Alpha.'

The service was led by the Reverend Andrew Wylie, chaplain to the oil industry, and as he began his address a woman sitting beside Punchard began to cry. He put his arm round her, and together they listened as Wylie spoke.

> God appears at such a time of bewilderment, in the sheer kindness of unspoken sympathy, the handshake, the arm around the shoulder, the hug, the understanding glance, the thoughtful gesture, the listening ear ... In recent years,

because Britain's offshore oil industry is over the horizon, it has been all too easy for those who work in it, through being out of sight, to become out of mind. This should not be. No matter the extraordinary level of technical achievement which makes complex exploration and difficult extraction possible, it is very much a people industry. And it is the people who are its most precious investment and give it its special character ... This part of the offshore family is no more – but the family remains and proudly goes on.

Mario Conti, the Catholic Bishop of Aberdeen, read from Psalm 46: 'God is our shelter and strength. So we will not be afraid even if the earth shakes and mountains fall into the ocean depths.'

Bob Ballantyne did not wish to attend the memorial service. He could not bear, he said, to sit in the same company as Occidental and the politicians on whom he now pinned the blame. The past few weeks had been a busy round of funerals for friends. The body of Charlie McLaughlin was recovered from the water two days after the disaster, and Bob and Pat drove to Glasgow for the funeral. The casket was open at the service and when Bob leaned over to pay his last respects, he smiled. Charlie's wife had insisted her husband be given a perm, as his hair had always been worn. 'A bear with a perm' was how Ballantyne described him.

The psychological effect of his ordeal was apparent to all. Unlike other survivors who were reluctant to discuss what they had gone through, Ballantyne would talk and talk and talk, but each time he told the story his terror returned. He would become agitated and frightened and his eyes would fill with tears. He was unable to sleep, and when he did doze off he found himself among flames and dark oily water. Then he would wake drenched in sweat and cling to Pat for company. He became an abysmally bad driver, unable to concentrate and disturbed by red lights. Yet within a fortnight he had renewed contact with his son and former wife, and arranged to be reunited with his daughter Amanda.

The vow made amid the water and flames was fulfilled under the birch and oak trees of Chatelherault country estate in Lanarkshire, near the family home in East Kilbride. Amanda, now sixteen, had recently left school and was working in a shoe-shop. It was six years since they had last met, and when Ballantyne held her he cried. They both did. In between the tears he assured her that, despite the burns and bruises, he was fine. He kept the nightmares to himself.

On the day of the memorial service, Ballantyne wished to pay his respects in private. He decided to cast wild flowers out to sea, but in his confused state could not decide on which beach to begin his ceremony. Beautiful coastline and sandy bays abound in Aberdeenshire, but somehow he wound up picking his way through a military firing range to reach the MOD's private beach and the sea.

Each survivor handled their experiences in a different way. Michael Jennings was one of the few who immediately decided to return to work and, one week later, flew out to the Claymore platform. His return was not without incident, however. At night he would go up onto the helideck and look towards the horizon and the ruined stump of Piper. One evening a man came up behind him and as he turned and smiled Jennings said: 'What a waste, eh!' He turned to find he was talking to himself. Michael Jennings was literally seeing ghosts.

•

The task of caring for the survivors and the families of the dead fell to a small group of social workers who took over a converted garage in Justicemill Lane, at the top of Union Street in

Aberdeen. A few sofas, potted plants and attentive ears created a safe refuge. The Piper Alpha Outreach office was set up in the days after the disaster by Anne Bone and David Tumelty, two senior members of staff with Grampian social work department, who would devote the next eighteen months to constructing the necessary emotional support for those on the brink of collapse.

The help-line, set up in the first 24 hours and designed to gather together accurate information from Occidental, the various contractors and the police, received hundreds of calls in the first few days. A letter was then sent out to both survivors and the families of the missing, explaining that staff were available to assist and enclosing a leaflet entitled 'Coping with a major personal crisis', which had been adapted from similar documents prepared after the Bradford fire disaster and the sinking of the *Herald of Free Enterprise*. It was a simple, practical idea and it was effective. A few days later one woman whose husband was missing called to request another copy as the first was being kept under his pillow by her 13-year-old son. The help-line's letter also explained that a follow-up would arrive within ten days, and that if they wished for a visit from a social worker one would be arranged, an offer which was accepted by 85 per cent of the cases involved.

The home visits were to prove beneficial to families but stressful to staff, as wives and girlfriends tumbled through various emotions including anger and rage at Occidental. For those whose loved ones were missing, a degree of fantasy crept into the conversations. A pattern emerged where wives, fathers and mothers concocted elaborate scenarios in which their loved one survived with amnesia and was picked up by a foreign trawler. Although rationally they knew this to be false, it was a spar to which they clung. When a body was recovered – as a number were in the first few weeks – the fantasy could still persist because the families were discouraged from viewing the body, which was delivered to the undertaker in a sealed coffin. This practice was to trigger a debate, with Anne Bone arguing that it was essential for the grieving process that the family see the body, while David Tumelty sided with the police that, depending on its condition, it might make matters worse. The situation was finally settled when social workers visited the temporary mortuary at Aberdeen airport, viewed a body and agreed that it was an ordeal the family should be spared. Yet, months later, one widow commented: 'How can I be sure it was him in that coffin?'

Over time a matrix of support meetings was set up, each attuned to the needs of their members. A number of widows' groups were spread across the north-east, while a parents' group was also formed for those who had lost their sons. Many of them felt they were 'the forgotten people' as the public concentrated on the plight of the widows and their young children. Yet they had to endure their own specific problems. While many proved to be a source of comfort to their daughters-in-law, according to a social work report some became rivalrous, particularly if their son had married only recently. The Piper Outreach Team also set up the *Piper Line*, a small newsletter distributed to everyone known to have a connection with Piper Alpha, which rose from a circulation of 300 to over 600 copies. The most poignant pages were the children's drawings of stick men amid a structure of flaming crayon red.

To reunite families with the body of a brother, a husband or a son was a battle against the elements and at the edge of engineering technology. While all four accommodation modules were located, only two were considered safe enough to raise and even they almost cost a diver's life. The East Replacement Quarters (ERQ) was found lying off the east side of the platform, in an area relatively clear of debris, while the Additional Accommodation West (AAW) lay to the west in an area similarly clear.

Over a five-week period between 8 August and 13 September, the contractors, Subsea, fitted on to the module a series of clamps and bolts which were then threaded with slings. When the initial lift began the ERQ, which was partially embedded into the seabed, rose up, but then a sling snapped and the whole module rotated vertically 90 degrees before falling back onto the seabed again. The module was eventually lifted off the seabed and lowered onto a metal frame on 4 October, but it was then discovered that a walkway on the module had not been accounted for in the calculations and knocked off the frame's centre of gravity. This necessitated additional rigging to hold it in place and prevent rotation as it rose from the water. The ERQ was finally raised by a derrick barge crane, a DB102, on 15 October, a lift that began just after midnight and took 20 hours to complete.

The AAW was undergoing simultaneous preparations, but as it was welded on to another module, divers had to separate the two before lifting could commence. To achieve this divers used an Oxy-Arc cutting with a Broco rod, through which oxygen is fed, but this carried the risk of a blowback caused by bubbles of gas or oxygen that had accumulated in the gap between the modules. While two minor blowbacks caused tingling to the diver's fingers, on 27 September a diver experienced a major blowback which cracked his helmet, perforated his eardrums and damaged his sinuses. He was dragged back to the diving bell, close to death and with his mask full of blood. Mouth-to-mouth resuscitation helped to save his life, but as he was diving at a depth of over 400 feet, and could not leave the decompression chamber on-board the support vessel, a doctor and nurse were flown offshore to treat him. The AAW was finally raised on 11 October.

An immediate search of the AAW revealed no bodies, while

foul weather prevented a preliminary search of the ERQ. Both modules were then taken to the Flotta oil terminal at Scapa Flow for the arduous process of recovering the dead. The task of searching piece by piece through the four-storey ERQ was the responsibility of Grampian Police who, while not shirking the duty, were concerned about the consequence to their officers' mental health. Inspector Gordon had previously worked with Dr David Alexander, at the department of mental health at the University of Aberdeen, on an extensive study of occupational health and stress within the Force, and now he invited his suggestions on how best to manage the operation. Dr Alexander aimed to reduce adverse reactions by taking the officers' requirements seriously, developing an esprit de corps among the team and persistently driving home the vital importance of their work to the families who still waited for the phone to ring with the news of their loved ones' recovery.

Twenty police officers, divided into four teams of five, rotated their duties, meaning that body retrieval and handling was limited to every third or fourth day. They were joined by another fifty support staff including divers, staff from Occidental and Kenyons, as well as people from the Department of Energy and the Health and Safety Executive.

The search was difficult. On each of the four floors the internal structures had collapsed with only two rooms recognizable – the galley, where the majority of the bodies lay, and a small room on A level which contained switchgear. As the module was upside-down, what had been the floor was now the ceiling, which meant that officers stepped carefully through sodden, blackened debris stacked to a height of several feet. The four floors remained connected by a passable stairwell which, along with steel bars that appeared at regular intervals on each floor, became the central points of reference on a grid system that was drawn up to plot the exact location of each body. The stench was primarily that of salt water and the large quantity of rotten food from the galley.

Concerns for health and safety meant that all those who entered the module had to wear protective clothing, consequently portacabins were erected and designated as 'clean' and 'dirty' areas. Officers came in through the 'clean' area, changed into thermal underwear, boiler suits and waterproofs and then accessed the site. Each day started at 8 a.m. with a detailed briefing on the specific target area, then continued – with a break at noon for soup and a sandwich in the kitchen area – until 6 p.m. As they left the module the teams entered the 'dirty' area where they discarded their boiler suits and underwear. They then entered the showers where clean clothes were laid out for them. After dinner the various team leaders met and under the chairmanship of Chief Inspector Gordon discussed the day's work and any issues that had been raised by staff, and planned the work schedule for the following day. The other team members unwound in a small bar for Flotta employees. During his frequents visits to the site, Dr Alexander sat in on the meetings and also mixed with team members in the bar so that he could quietly assess their stability. The search took five weeks to complete.

The conditions of the bodies, submerged for twelve weeks, varied dependent on the trauma inflicted during the collapse. Head wounds were common. Each officer had their own coping mechanism for the sight. 'Sir, when I go in there, as far as I am concerned I'm going into a spaceship looking for Martians,' explained one senior detective to Dr Alexander. While this man erected a wall between the bodies and their previous humanity, others embraced it, talking to those they carried as to a fallen comrade: 'Right, laddie, let's get you home.'

Once each body had been carefully lifted from the module,

with people pausing in respect as he passed, it was checked for any sign – such as a tattoo – that could link it to an individual. Then the remains were placed into a hermetically sealed aluminium coffin and flown back to the principal mortuary at Aberdeen airport. The body was then checked against the log of personal details already acquired on each of the missing, including hair samples and fingerprints which had been gathered from their home. The officers at Flotta were inspired and delighted every time another person was identified, and for some the work took on the role of a personal mission. A research paper by Dr Alexander would later reveal that officers suffered no adverse stress as a result of their duties.

For the families of the missing, the autumn was a period when every phone call triggered a reflex response that 'his' body was coming home. Molly Pearston found the tension to be almost unbearable. She had returned to work at Tornadee Hospital where the staff supported her as family. Each time the ward sister came to say that there was a phone call, Molly could not help but recognize the same hope in the sister's eyes, but it would not be rewarded. Every call was about something else. When the bodies of Robert's fellow crewmen were returned to their families, Molly attended the funerals. Mark Ashton was just nineteen, a trainee technician with Macnamee Services, who died of smoke and gas inhalation in the galley, from where he was recovered on 24 October. Sitting in the church pew at his funeral service she felt a faint kind of connection, but it was not the same. The loss of her son was a physical pain that reignited each day at dawn when she awoke from dreams of Robert as a child. He was always a boy when she slept, never the man he went on to become. It was as if his vulnerability had returned. In dreams – nightmares some of them – he was always in need, the gentlest of which focused on his requirement for new Clarks shoes or a pair of school trousers.

Across the Atlantic, the long wait endured by the Busse family began to draw to a close when, at 12.35 p.m., on 24 October, another body was removed from level D of the ERQ, just to the east of the stairwell. Carl William Busse was later identified by his teeth, survival suit and jewellery, and was released back into the care of his family on 1 November. Like so many others, he had died of smoke inhalation.

The Busse family found a measure of closure in bringing their brother home to the Salem Lutheran Church where he had worshipped and where Janis, his sister, taught Sunday school. He was buried under a beech tree in the graveyard on 8 November in a quiet family service. A larger memorial service had already taken place in July at the First Baptist Church in Navasota, where Carl's wife worshipped, and there were plans to set up a scholarship in his name.

Yet the comfort of Carl's return could not dispel the questions Janis and the family would ask themselves repeatedly: How was he that day? Did he look worried? Had his faith brought him comfort? It would be years before Janis could bear to look at her brother's picture.

18. PAYING FOR THE PIPER

The legal response to Piper Alpha began at about 7 a.m. on Thursday 7 July. As the last of the helicopters bearing the injured touched down on the helipad at the Royal Infirmary, David Burnside, a local solicitor, picked up the telephone ringing by his bed. He had retired early the previous night and so was unaware of what had happened. The caller, Roger Pannone, a solicitor from Manchester, briefed him in preparation for the many calls from the media that soon followed.

The two men had first met two years previously when Burnside formed a syndicate of solicitors to represent the families of the 45 victims of the Chinook disaster. Pannone, who had pioneered the idea in response to the 1985 crash of a British Airtours flight at Manchester airport in which 55 people were killed, had visited Aberdeen to offer advice and support. He was aware that as Burnside was the press spokesman for the Chinook families, the media would contact him for comment on the legal responsibility now faced by Occidental. Both men were shocked by the scale of the disaster, but both knew that legal action was inevitable and that a large collective of victims would be able to put the most pressure on Occidental. The two men then discussed grounds for action in America and, most importantly, in which state. Burnside had experience of pursuing Boeing Vertol, the manufacturers of the Chinook helicopter, in Philadelphia, but different states had different attitudes towards outside litigation.

Pannone explained that while Occidental's headquarters were in Los Angeles, California had a law that did not look favourably on actions raised in their courts for incidents that had occurred elsewhere. Texas and Louisiana, where Occidental also had prominent offices, were considered far more promising.

Occidental, as it would turn out, were keen to settle but not at any price. The words spoken by Dr Hammer at the Aberdeen press conference, where he assured families that the company was well insured and that everyone would be looked after, provided the syndicate of lawyers acting on behalf of both the families of the dead and the survivors with a solid advantage, one that allowed complex negotiations to be contracted into months instead of years. Although Occidental's staff on board numbered just 23, it was agreed with the other contractors that they would lead the negotiations, backed by the company's insurers, Lloyd's of London, with the contractors' liability calculated afterwards. As David Burnside saw it, Occidental had two reasons for haste: First, a swift, generous settlement would cast the company in a brighter light ahead of the public inquiry. Second, it would prevent Scottish lawyers – who had made clear their intention to set sail – from docking in an American court where compensation claims were, on average, ten times more generous than those provided under Scottish law.

Within one week of the disaster the Piper Alpha Disaster Group (as it would come to be known) had appointed a steering committee of eleven elected members and had already held an initial informal meeting with Chuck Foster, Occidental's in-house counsel, who had flown over from the company's headquarters on Wilshire Boulevard in Los Angeles. This meeting took place over coffee in the boardroom of Paul & Williamson, the Aberdeen law firm appointed by Occidental to represent them. Tradition dictates that a lawyer keep his emotions locked in a filing cabinet, but Foster, a quiet pipe-smoker, was visibly upset by the circumstances of his visit and repeated the words of Armand Hammer that everyone would be fully compensated. The question was, at what level?

The Scottish lawyers felt an onus to their clients, who were growing by the day, to secure a swift and generous settlement; but while legal proceedings in America offered the potential of a greater sum, this came with the guarantee of a lengthier wait. Instead, the idea formed to secure a 'mid-Atlantic settlement' – a formula that would generate a sum midway between that provided by either nation. Over the forthcoming months of negotiations, the lawyers pushed for the settlement figures to be closer to Manhattan rather than Stranraer. While the group would eventually come to represent 95 per cent of the families and survivors, the complexities of the case multiplied, particularly when the deceased had been married more than once, and where different solicitors represented each wife.

Not everyone chose the Piper Alpha Disaster Group to represent their interests. In May 1988 Frank Lafevre, a prominent Aberdeen solicitor, had angered the legal establishment by setting up Quantum Claims, Scotland's first 'no win, no fee' legal firm, which was seen as a challenge to the solicitors' current cartel. The disaster was to provide sixteen cases for Quantum Claims and allow Occidental to try setting a precedent for a suitable level of compensation. While Paul & Williamson were meeting with the Piper Alpha legal team, they were also juggling multiple negotiations with a clutch of different law firms who had their own clients. Lesley Gray, the senior corporate lawyer at Paul & Williamson charged with handling the negotiations, saw the benefit of settling swiftly with one of Lafevre's clients, a widow with two young children. Lafevre was in the bath when Gray called to tell him that she wished him to be the first to achieve a settlement. The figure was £980,000, of which Quantum Claims received £70,000. Not all Lafevre's cases would proceed as smoothly, but one that did illustrate the benefit of persistence and patience was that of a diver who, dissatisfied with his initial offer, took his case to mediation in Texas and achieved a settlement of $875,000.

Meanwhile the Piper Alpha Disaster Group continued to meet in the offices of Paul & Williamson, where the boardroom had a table which was long enough to accommodate both sides. While Occidental were leading the negotiations, various contractors also wished to be represented. The Disaster Group's 'ace in the hole' was Stuart Speiser, an American veteran of such civil cases who flew over frequently to advise the group, while remaining in the background, and did not attend the actual negotiations. Despite Occidental's initial goodwill, talks ground to a halt at times as the figures escalated, but Burnside's frequent quotations from the multi-million payouts instructed by the US courts would provide the necessary jolt.

The discussions centred around constructing a formula that could be applied to each individual who had died, taking into consideration their seniority, current salary, future potential earnings and number of dependants. Broad strokes were agreed within 100 days, but the fine details took over a year to complete. Part of the delay was over the 'release and discharge' clauses on the final documents, which according to one reading appeared to grant Occidental the right to claim back compensation payments at a later date if new information came to light. The clause was suitably amended and the final agreement was concluded in November 1988 at a high-powered summit in the offices of Roger Pannone in Manchester. The paperwork was signed at 9 p.m., after which both sides ordered a Chinese take-away. David Burnside was unable to linger as he had an appointment back in Aberdeen. Anyone who suspected that the group's media spokesman harboured a love of the spotlight would be unsurprised to discover that he was playing the role of Widow Twanky in his amateur dramatic society's pantomime.

The monetary value of the lives lost on Piper Alpha ranged from the highest figure of £1.3 million to an average of £700,000. The lowest figure paid was to the parents of young single men, whose lives were calculated to be worth between £150–£200,000.

The negotiations concerning the survivors dragged on longer, with many having to endure a battery of physical examinations and in-depth interviews to establish the extent of their mental trauma. Those who bore the visible physical scars from burns found it easier to prove their case for adequate compensation than those physically unharmed, but who woke each night screaming in sheets soaked in sweat. Tragically, in the end one man would consider his life to be worthless. Dick Common, the diver's clerk, could never accept why he – a single man without children – should have survived while his boss, Barry Barber, Occidental's diving consultant and a family man, had drowned. He would often call Bob Ballantyne and talk for hours about why he had survived. Ballantyne's repeated attempts to convince him of his own luck and good fortune would ultimately prove unsuccessful. Unemployed since the disaster, Dick Common cut his wrists; his body was discovered in his flat in Aberdeen on 12 August 1994. Doreen Jennings, the wife of Michael Jennings, was also a recipient of Common's late-night phone calls and said at the time: 'He died because of Piper Alpha. It never left his mind. It was like a nightmare that went on and on.'

The final bill to compensate the families of the dead and the survivors would top £110 million, from which dozens of solicitors and lawyers collected a share. The fee payable to the legal establishment was calculated by using what was known as a 'chapter 10' fee, a formula proposed by the Law Society of Scotland. In effect, if a compensation claim was 'x', the lawyer should be paid 'y'. In an attempt to ensure that as much money went to their clients as possible the formula was designed to achieve a sum equivalent to 'x' plus 'y'. While not Herculean, the legal work for those directly involved was exceptionally heavy.

The cheques began to arrive after the first anniversary of the disaster. For Joe Meanen, who required seven operations on his burned arms, the money took him around the world. He was a young man who had earned a second chance at life, and he decided to live it to the full before eventually returning to his home town where he bought and still manages a local pub. Ed Punchard used the money to move from the cold waters of the North Sea to Australia and the shark-infested waters of the television industry, where he became a successful film-maker, winning awards for a personal film about the disaster which was broadcast on the tenth anniversary. Bob Ballantyne, as his friends and family attest, was 'screwed'. Ballantyne, a proud, strong-willed man, had underplayed his suffering and received a sum which, when compared with those who benefited from a swifter, easier escape, was paltry. The sums for survivors ranged from £30,000 to £1 million. One survivor refused to accept a penny, such was his contempt for Occidental.

19. THE CULLEN INQUIRY

The Public Inquiry into what took place on board Piper Alpha to result in the deaths of 167 men, took thirteen months, running between January 1989 and February 1990. It called 260 witnesses and throughout its length a total of six million words were spoken. The survivors told of their escape, often accompanied in the witness box by their wives who held their hands. Short breaks to comfort the distressed were common. Lord Cullen, the High Court Judge who led the inquiry, listened to every word and his two-volume report, *An Inquiry into the Piper Alpha Disaster*, was published on 12 November 1990. It penetrated far beyond what one trade union official had claimed prior to the inquiry would be to simply ask 'the narrow technical question of whether a valve worked or did not work'.

The report was damning, in effect an indictment of a culture of complacency at Occidental where the monitoring of work was inadequate in an environment where mistakes proved lethal. The permit-to-work system was 'knowingly and flagrantly disregarded', relying on 'informal communication' between personnel instead of the strict observance of proper procedure. Lord Cullen also found there was 'no formal training in the permit-to-work system'. Occidental left the responsibility for training workers to their contractors. The permit-to-work system was supposed to be monitored and audited, but this had not been done in the twelve months prior to the disaster.

Occidental's attitude to the 1987 report that specifically raised the problem of a ruptured gas riser and a prolonged high-pressure fire was found to be blinkered. Lord Cullen wrote:

> I must make every effort to avoid being influenced by hindsight, but making all allowances for that I consider that management were remiss in not enquiring further into the risks of a rupture of one of the gas risers and in such an event the risk of structural damage and injury to personnel.

Occidental's assessment of risk was considered unsatisfactory, while the ability of management to review and monitor safety procedures was lacking. They failed to perceive how changes in equipment and activities had serious safety implications. The decision to maintain oil and gas production during a massive period of ongoing construction and crucial maintenance work was described by Lord Cullen as 'puzzling'. This is a word that, like an usher in a theatre, silently points the reader in only one direction: that perhaps Occidental weighed up the risk versus the reward and found the bottom line to be a better balance.

Lord Cullen wrote in his conclusion:

> It appears to me that there were significant flaws in the quality of Occidental's management of safety which affected the circumstances of the events of the disaster. Senior management were too easily satisfied that the PTW (permit-to-work) system was being operated correctly, relying on the absence of any feedback of problems as indicating that all was well. They failed to provide the training required to ensure that an effective PTW system was operated in practice. In the face of a known problem with the deluge system they did not become personally involved in probing the extent of the problem and what should be done to resolve it as soon as possible. They adopted a superficial response when issues of safety were raised by others, as for example

at the time of Mr Saldana's report and the Sutherland prosecution. They failed to ensure that emergency training was being provided as they intended. Platform personnel and management were not prepared for a major emergency as they should have been.

The *Daily Record*, on the day following publication of the Cullen Inquiry Report, carried two words on the front page: 'Charge Them!'

Yet there would be no criminal charges brought against any member of Occidental's management team. The Lord Advocate, Lord Fraser of Carmyllie, reviewed the reports from the procurator fiscal's office and in July 1991 announced in a letter to Frank Doran, the Labour MP for Aberdeen South, that he had decided that there would be no criminal proceedings on the basis that there was 'insufficient evidence'. It was uncommon for the Lord Advocate to give specific reasons for deciding against any prosecution but in this case it was explained that a successful criminal prosecution would require proof 'beyond reasonable doubt'. Unfortunately, as many of the key witnesses had been killed in the disaster, Lord Cullen's report had been based on what was described as 'inference'. This was a decision that provoked outrage from the Piper Alpha Families and Survivors Association, who for a time considered launching a private prosecution against the company on the grounds that whatever the deficiencies in the existing criminal negligence laws, there had clearly been breaches of health and safety legislation. However, it was reluctantly decided that this was beyond their financial means. It was felt that the government lacked an appetite for any criminal prosecution since they too had been required to stand beside Occidental in the 'dock' as contributors to the culture of complacency that had become common practice across the North Sea. At the time of the disaster there were 139

ONSHORE

fixed installations and 76 mobile rigs operating in the British section of the continental shelf. The number of government inspectors with responsibility for ensuring the safety of these 215 installations was five. The British government – in order to encourage the vast investment required in the North Sea – had historically been generous to companies who in the early days operated, literally, beyond the law, with current health and safety legislation such as the Factories Act only applicable on-shore. Yet even when the Health and Safety at Work Act (1974) was introduced, it took until 1977 for it to become applicable offshore and even then the inspectors were drawn not from the Health and Safety Executive but from the Department of Energy. Although the inspectors were required to investigate all fatalities and accidents that involved extensive injuries, just 40 per cent were actually investigated, in principle to discover what safety lessons could be learned, but in practice they were useless. One month prior to the Piper Alpha disaster a government inspector had visited the platform in the wake of the Sutherland fatality; he spent ten hours touring the platform, after which he declared himself satisfied that the handover between shifts had been 'tidied up'. Lord Cullen described the visit as 'superficial to the point of being of little use as a test of safety' and dismissed the inspectors as 'inadequately trained, guided and led' and whose 'persistent under-manning' affected both the frequency and the depth of their investigations.

A total of 106 recommendations were made by Lord Cullen in order to increase the safety of offshore work. At the heart of these lay what became known as a 'Safety Case', a document that required the operator to review every potential hazard on an installation and construct a management safety system to deal with each one. He wrote: 'The offshore Safety Case ... should be a demonstration that the hazards of the installation have been identified and assessed, and are under control, and that the exposure of personnel to these hazards has been minimized.' A central plank in the Safety Case was that each installation be equipped with a 'temporary safe refuge' (TSR) which was capable of withstanding the heat and flame of an inferno long enough to allow workers to evacuate safely.

On the day when the Cullen Report was published, Glenn Shurtz, president of Occidental Petroleum (Caledonia), insisted: 'We have always practised the management of safety. Offshore it's our number one priority.' At a press conference Shurtz attempted to fend off any criticism by insisting that he could not respond to criticism he had not read: 'We have just received Lord Cullen's report, and it is a little unfair for me to accept criticism which I haven't had a chance to look at.' Nevertheless, the company insisted that they had already spent £50 million on improving safety, including £28 million on emergency shutdown valves in the Claymore field.

Dr Armand Hammer died a month after the publication of the Cullen Report. He was 92.

20. COGS IN A BROKEN MACHINE

The legacy of Occidental's 'swift' settlement would result in the longest-running legal case in Scottish history, with a senior judge pointing the finger of blame at two dead men.

Occidental's agreement to negotiate and pay compensation to families and survivors was on two grounds – that the company did not accept liability, and that it would seek to recoup the sum paid from each of the 26 contractors with staff on-board. The reason was simple. Occidental's contracts with each company contained arrangements and indemnities that they believed entitled them to reclaim from the company the sum paid out to their specific employees. While Occidental had been required to pay a proportion of the £110 million from their own reserves, the bulk was paid by Lloyd's of London, the company's insurers, who also had to pay £2 billion for the construction of a replacement platform, Piper Bravo. Lloyds had endured a run of multi-billion payouts in 1988, including cases arising from damage wreaked by Hurricane Andrew in Florida and Hurricane Hugo in Puerto Rico and Carolina. Therefore they were intent, where possible, to claim back money where they could, in this case from the insurance companies behind each contractor.

The decision to take legal action was made in 1988, with the first legal papers served the following year, while Lord Cullen was still carrying out his inquiry. The threat of legal action was successful in one case. Noble Drilling, previously Bawden International, settled out of court by making an undisclosed payment, yet the remainder of the contractors held firm. The publication of the Cullen Report served only to strengthen their resolve and their belief that Occidental alone was responsible for the disaster. The report had been highly critical of Occidental management, but also traced the source of the explosion to the flange put in place by employees from Score UK Ltd, the one contractor who was not facing legal action (though Occidental reserved the right to take such action at a later date). Score UK Ltd was omitted from the lawsuit in order to allow Occidental's lawyers to focus on the legal agreements with the contractors as well as perhaps depriving opposing counsel of a suitable scapegoat.

Occidental left the North Sea in 1991. Occidental Caledonia was sold to Elf Aquitaine for $1.35 billion in cash, with the French company also assuming $130 million of debt. The deal, part of a global restructuring strategy that pushed Occidental back into the black, also included an extraordinary tax credit valued at $112 million, worth half the estimated value of the lives lost and those left psychologically scarred. In Occidental's 1991 annual report no mention was made of Piper Alpha, but the net gain on the sale of its North Sea assets was $646 million. When Elf took over the company, they also inherited the lawsuit against the contractors.

In the scales of justice at the Court of Session in Edinburgh, the conclusions of Lord Cullen carried all the weight of just one man. The civil case against the contractors, which opened on 1 March 1993 in front of Lord Caplan, would in effect be a retrial of arguments as to the disaster's cause with some new theories put before the court. Alex Rankin, who with Terry Sutton had removed PSV 504 and who was served with a 'notice of blame' by Occidental during the Cullen Inquiry, was compelled to testify. Although described as an 'unreliable witness' by Lord Cullen, Rankin had later insisted that he was intimidated by the inquiry and unsure of his legal position. He testified that it was the corporate system, not any individual, that was to blame.

The civil case, as everyone involved admitted, was about money: £140 million, once Occidental's legal fees and interest had been included. But one man provided a regular testimony to the human cost. Gavin Cleland, a former miner and retired care-home worker, lost his son Robert, a derrickman, in the disaster. Cleland was consumed, like no other relative, by the desire to see Occidental executives charged with criminal negligence. A small man who wore thick black glasses, he carried a home-made banner which featured the Piper Alpha platform ablaze with flames of red silk. His campaign to see Occidental in the dock involved a relentless letter-writing campaign to Scottish MPs and the judiciary. Few responded, those who did so offering sympathetic words but little support. Like many parents, he was haunted by what had happened to his son during that night. All that was known for certain was that Robert was on the night shift at the time of the first explosion and that his body was recovered from the water on 10 July with the cause of death attributed to inhalation of smoke and gas. Cleland, a lifelong communist, believed that money was the reason for the government's failure to prosecute. The compensation payments had helped to sap the public's support for criminal proceedings, on the grounds that the families of the victims were now financially secure.

The hearing that began on 1 March 1993 ended on 31 October 1996, having lasted 391 days, almost twice as long as the previous record of 201 days. In total 13 million words were spoken, 64,515 pages of evidence examined at a cost of £10 million in combined legal fees.

The death of a princess on 31 August 1997 meant that the publication of Lord Caplan's findings on Monday 2 September attracted little coverage. In the test case against seven contractors, he found against Elf (Occidental) on the grounds that two workers, one of whom was employed by Occidental, had been negligent. The named men were Terrance Sutton, who had fitted the flange with finger-tight screws, and Robert Vernon, who had started up Pump B which triggered the initial explosion.

In his judgment he wrote: 'I am well aware how unfortunate it is that I blame Mr Vernon and Mr Sutton who both were killed by the accident and therefore are not able to explain and possibly exonerate themselves.'

Nonetheless, Lord Caplan went on to insist that Terrance Sutton had been negligent in failing to properly tighten the screws on the blank flange. He wrote: 'There is the inescapable inference that he failed to fit the flange properly. The most likely cause of the leak was that he had failed for some reason to tighten the securing bolts in the manner he had been instructed and which he knew to be the correct method.'

On Robert Vernon's actions, he wrote:

I am led to the view that there is a marked probability that this accident was caused because Mr Vernon proceeded to introduce hydrocarbon to the pump at the time when this should have been avoided – when [the valve] was not in place. He did not know that the valve was not in place [but] as lead production operator he should have taken care to know what was going on within production processes on the platform at any time and not to make mistakes in that regard.

He then wrote:

Thus the accident was caused by negligence on the part of Mr Vernon and Mr Sutton.

The Law Lord found that in the circumstances Occidental (now Elf) had no option but to settle the claims, and that the terms of the settlements were reasonable. As the insurers, Lloyd's, had paid the compensation, Occidental had not suffered any loss and, as a result, the contractors were not bound to indemnify the company. Elf was required to pay legal fees of £10 million. The decision to embark on a court case that would result in the pinning of blame on the backs of two dead men, who were but cogs in a broken machine, did not leave Occidental's insurers empty-handed. A slight difference in the written contract with one of the seven contractors meant that Elf were entitled to receive £12,685.

The cost to the families of Robert Vernon* and Terrance Sutton was considerably higher.

---

* Robert Vernon was posthumously awarded the Queen's Commendation for Brave Conduct, one of 20 awards bestowed for courage on the night of the disaster.

EPILOGUE

The man behind the desk in Occidental's headquarters in London made Sue Jane Taylor an offer the artist would eventually refuse. This was a few weeks after she turned on the radio in the small studio in Hackney – where she was at work for her forthcoming exhibition, Oil Worker Scotland – and learned of the disaster that had befallen the platform on which she had photographed and sketched. It took a few minutes for her to realize that many of the men who had shown her such kindness might now be dead, a fact later confirmed when she learned that neither Colin Seaton nor Gareth Watkin were among the list of survivors. Life on Piper Alpha now existed only in the acrylic and Conté sketches that hung on her studio walls. This fact was not forgotten by the public relations department at Occidental, who wrote to invite her to a meeting to which she was requested to kindly bring all photographs, sketches and paintings of Piper.

When she attended, however, she decided to leave her work behind. The public relations man, whom she had met several times before, was clearly under considerable stress. Fidgeting, furrowing his brow and smiling in a manner that conveyed anything but happiness or wellbeing, he quickly came to the point. Occidental wished to buy up every piece of art inspired by her visit to Piper Alpha, as well as the negatives of all photographs taken while on board. She could name her own price.

Taylor was taken aback and almost as nervous as her host. Since the accident she had been concerned about the morality of going ahead with her exhibition, which was due to open at the City Art Centre in Edinburgh in the autumn. Yet she felt a greater concern about capitulating to the deep pockets of a multinational oil company which was clearly intent on gathering up any evidence that might be detrimental in the forthcoming inquiry. She also felt the offer was a slap in the face for every oil worker she sought to celebrate through her art.

'I need time to think it over,' she explained before excusing herself. In truth, she had no intention of selling out, but was anxious to take legal advice on whether she could be compelled to hand over the work. She also wished to speak to the families of the men and survivors about their attitude towards the exhibition. As it happens, she was due to return to Aberdeen to complete a final series of prints at the city's Peacock Printmakers workshop, and so arranged to pay a visit to the Piper Alpha Outreach Team. While there, she talked through her own feelings of sadness with a qualified therapist and was introduced to Bob Ballantyne, who had been trying to exorcize his own experiences through paint. When he lifted up a painting to show her, it depicted nothing but burning fires stretching out to each corner.

Ballantyne told Taylor that he had noticed her during her visit, but had decided not to speak to her on the grounds that she was probably 'one of Oxy's people' and not to be trusted. She discussed her concerns, but he was supportive and helped to arrange a larger meeting with a mixture of survivors and relatives at which she could show her work. The outcome was positive. In the relatives' minds, the artwork was part of a visual memorial to the men who had died, and deserved the widest possible exposure. Suitably reassured, Taylor wrote back to Occidental (having first checked the wording with a lawyer) turning down the company's offer and explaining that the exhibition would go ahead as planned.

The opening night would bring a convoy of buses carrying relatives and survivors from Aberdeen to the City Art Centre in Edinburgh where they applauded the paintings that would haunt them. In 'Bond Shop Man' the moustached supplier of duty-free cigarettes, cigars and perfumes stared out of the frame, unsmiling, like a ghost conjured from Conté and charcoal. The 'Piper Alpha Crane', in pen and ink wash with pastel, was a cheerful image in bright yellow until the eye was caught by the orange flame of the gas flare at the side. 'Rigger II' was a colour etching of a smiling man with a smoke-grey face and a suit of flaming red, while 'John the Electrician', a thin narrow painting, resembled a medieval saint beneath a celestial white sky. All the paintings were completed before the disaster, but none could be viewed except through a prism of smoke and flame. The exhibition would go on to garner critical acclaim and eventually tour twelve venues across Scotland.

•

When the bereaved families decided that they wished to erect a permanent memorial to the dead, Occidental was discouraging. To the company it was enough to have a leather-bound book of remembrance placed at Aberdeen City Art Gallery, but others vehemently disagreed. Among the most persistent was Dr Sutherland, the father of Stuart Sutherland, the 21-year-old student who was working on Piper Alpha as a cleaner and whose body was never recovered. Among the relatives, Dr Sutherland had the clearest idea of what the memorial might resemble. He argued, persuasively, that it should be figurative as opposed to interpretive and might be modelled on the memorial at Spean Bridge in the Highlands to the Commandos who trained in Scotland during the Second World War. The plinth bears the motto 'United We Conquer' and features three larger-than-life soldiers, each facing in a different direction.

To Sue Jane Taylor it was a commission she was anxious to secure and felt fated to complete. While other artists were also approached for their ideas, which would focus on three figures symbolizing Piper's offshore workforce, Taylor's sketches of men in a moving series of configurations, combined with her personal connection to the platform, secured her the weighty responsibility. While she set about creating a maquette model that required the approval of the arts and recreation department of Aberdeen City Council, the bereaved families' memorial committee was struggling to reach their target of £100,000. Occidental refused numerous requests for assistance, including a letter from Bob Ballantyne suggesting that they donate the scrap value of the accommodation module from which the majority of the dead were recovered. The company was also putting pressure on other oil companies not to participate. Seven of those approached failed to reply to the memorial committee, while the remainder contributed a total of just £14,000 with individual sums ranging from a maximum of £2,000 down to just £150. The Scottish Office would eventually step in to cover a shortfall of £40,000.

Once the maquette had been approved, Taylor moved into the Scottish Sculpture Workshop, which was a collection of large concrete-block sheds in the village of Lumsden in Aberdeenshire. The isolation provided by the surrounding countryside was perfect for Sue Jane, who decided that the only way to have the sculptures ready for the unveiling in nine months' time – including three months spent in casting – was to dedicate herself with monastic rigour. Her 'cell' was a large outdoor shed; inside its unrendered breeze-block wall, through which dripped rainwater and snow, she created a protective cocoon by stapling bubble wrap over the draughty, creaking wooden doors.

Three models were required to pose for the figures. While two were arts graduates, it was the third who brought the greatest emotional resonance to the role. Taylor was keen to use an ex-oil worker, but unfortunately the first volunteer was unable to cope with the long periods of standing. She was then introduced to Bill Barron, who had been on the platform that night and whose wife, Trisha, believed he might benefit from the experience.

Like other survivors, Bill Barron was troubled. Unable to speak of what he had gone through, he sought instead to drown it with bottles of whisky. The pub had became a more tolerable place than home, and when he was at home Trisha did not know how his dark mood would manifest itself. One evening she returned from work to find him standing at the bottom of a 6ft hole he had dug in the garden, on a whim. He could not explain why, but it seemed to resemble a grave. Bill always kept a £1 note in his wallet as a visual reminder. After the accommodation block was lifted his wallet had been returned, containing one £20 note and three singles. He framed the £20 note, kept a single for himself and gave one to his brother. When he did think of 6 July, he often remembered the fish in the tank in the galley and wondered if they had survived the platform's collapse into the sea, if the death of so many had brought them liberty.

A small man of 55, with a thick moustache and a quiet manner, Barron had completed his National Service with the Black Watch before joining John Howard, a civil engineering company, where he worked on railway bridges and tunnels as a gaffer in charge of concreting work. He first went offshore in the late seventies when he joined the Wood Group. He spent the next ten years with them, most recently leading a paint squad 'chasing the rust' on Piper Alpha. Barron had just bought a new car and figured the visits out to Lumsden would 'be a good fine run' for him and his springer spaniel, Biff.

The three figures in the monument were to represent different roles. As Sue Jane Taylor would later write:

> The 'roustabout' bronze figure which faces west represents the physical nature of many offshore trades. His pose emphasizes two opposite 'strain' movements in offshore work: push and pull. On his right sleeve is a 'tree of life' motif, based on the Celtic design. Its leaves are in gold leaf. The design's mythological meanings symbolize the exploration and production of crude oil. For example, its roots deep in the bowels of the earth represent the vast oil wells underneath the seabed. The tips of its branches reaching up to the sky, the eternal flame of the flare boom on oil and gas production platforms; the vapour rising into the earth's atmosphere.
>
> The 'survival suit' bronze figure, which faces east, represents youth and eternal movement. On the left sleeve of this figure is a design of a sea eagle's wingspan and its head, gilt in gold leaf. The sea eagle is native to the northern seas, and it is used in place of the North American eagle, the patron of oil.
>
> The central bronze figure, which faces north, represents a mature character. In his left hand he holds a pool of oil sculpted in the shape of an unwinding natural spiral form. This black shape in his palm flows into gold leaf. His right hand points down to the ground, indicating the source of crude oil. The carved motif on his helmet, a fish and seabird design, symbolizes the environmental aspects of the oil industry's presence in the North Sea.

Bill Barron used his contacts to source a proper survival suit, overalls, boots and hard hat and, appropriately enough, took on the role of the central bronze figure. Each day he would drive out to the workshop, pull on his dark blue overalls, his steel-toe-capped boots and his hard hat and go to work, spending hours in fixed representation of a job he would never again do and men he would never again meet. While Bill stood on a wooden pallet, springer spaniel Biff sat by his side on an old worn doormat.

Taylor had began to work on the larger-scale models using clay; but finding it insufficiently flexible as well as requiring constant watering to keep it damp, she eventually abandoned it for the comfort of plaster. Wearing an orange boiler suit, flecked with white plaster powder, and with her hair tied back in a ponytail, she toiled through a cold winter that blanketed the area with snow and dull days that became darker still with the death of her 12-year-old niece Alison, who had died of peritonitis. Into a project already suffused with more grief than perhaps even bronze could bear, she now had her own to add.

At times the emotional pressure of dealing with visits from the bereaved, who would call in to inspect the work-in-progress – some of whom were unsupportive of what they saw – was almost unbearable. Yet she persisted in manipulating the rhythms and movements of the figures until they conveyed what she wished: presence, poignancy, symbolism and, most importantly, hope.

All through this time Bill Barron was a comforting presence. As she worked he would stand with his left hand outstretched and his right curled into a fist with one finger pointing to the ground, and they would talk. At first the conversation centred around the safe subjects of the past, his army days and time spent cleaning and painting the colossal cooling towers of England's power stations. Then, as the modelling sessions stacked up, he began to open up about the events of 6 July.

Like other survivors, Barron carried a hand-made cross constructed of guilt. Earlier that July evening, as the rest of his crew were returning to their accommodation on the *Tharos*, one lad who had recently married asked if he could have some overtime and so Barron found him a few hours' work preparing a section of the platform to be painted the following day. He never escaped.

Barron was billeted on-board the Piper and was in the cinema when the first explosion struck. Like others he had struggled to find a safe route out, but eventually made it onto the pipe deck where he discovered that much of the smoke – which had prevented many people from moving out – was coming from a burning skip. The route was passable, even if it didn't appear to be, yet when he returned to the accommodation module in an attempt to persuade others to follow, they refused, insisting that they had to wait for their supervisors to tell them the best way out. Barron repeatedly looked for his young crewman, with the certain knowledge that if he, as his gaffer, found the lad, he would certainly follow Bill's instructions.

It was clear that Barron found talking of that night exceedingly difficult, but pieces of his story would drip out with each visit. How he tried to inflate a life raft only to see it swept under the platform; and, unable to swim, how he dangled from a rope until an FRC eventually came along to collect him. Taylor, listening intently, would then press each scene into the plaster man who towered above her.

Once the moulds were complete, Sue Jane Taylor made repeated visits to the Burleighfield Foundry in High Wycombe during the three months it took to cast the statues. The plinth, made from Corrennie pink granite, chiselled with the names of each of the 167 men killed, was erected in the centre of a rose garden in a quiet corner of Hazelhead Park, on the outskirts of Aberdeen. Perhaps understandably, the completion of the statue coincided with a collapse in Taylor's health, as her work done, her body could now give in to the flu. Confined to her little room at the workshop, she had no choice but to stagger out to the lobby, shivering, as she tried to discuss preparations for the Memorial's unveiling.

The 6 July 1991 was a rare hot summer's day when 1,000 people gathered among the roses of Hazelhead Park to see the Queen Mother tug back the veil and reveal the finished piece, the men standing in perpetual vigil over the names carved in the stone below. Within the stone itself, the ashes of those unidentifiable remains recovered from the accommodation block had been placed, for this was more than a monument, it was a tomb. As the statue was revealed, workers offshore also held their own memorials. On the Brent Bravo there was an hour's stoppage in remembrance, while on the Safe Gothia extracts from the testimony of survivors to the Cullen Inquiry was read aloud along with poems of remembrance. According to the book *Paying for the Piper*, Press Offshore, a major contractor, permitted a fifteen-minute remembrance but docked the pay of anyone who took an hour. Occidental, meanwhile, posted a memo permitting a one-minute silence.

Among the many survivors attendant in the park stood Bob Ballantyne. He would graduate from Aberdeen University and would go on to teach and to raise two beloved daughters with Pat, but the flickering shadows cast by the fires of Piper Alpha would never leave him. Each anniversary for the rest of his life, he returned to the park to stand for ten minutes in silent contemplation beneath the bronze men he had left behind. Even in 2004, when cancer did what the flames could not, a piece of Piper Alpha was present at Bob Ballantyne's funeral when the minister read from the same dog-eared Penguin paperback edition of *Candide* by Voltaire.

For Sue Jane Taylor, back in Hazelhead Park on that summer's day, the emotion was almost too much. It was as if all her past experience and interests and talent had been a sea sweeping her to this point. Although buoyed by the kind comments from the gathered relatives and friends, she dearly wished to drift away.

•

On the first anniversary, Molly Pearston stepped on board the *St Sunniva*, the P. & O. ferry chartered to carry the families of the thirty missing men. She was accompanied by her younger son, and together they had brought a wreath of white flowers. Molly was so desperate to be reunited with Robert that she had visited a spiritualist church in Aberdeen. For a time, the woman's description of a young man in a sailor suit and big black boots had brought comfort.

The 120-mile journey from Aberdeen harbour took almost twelve hours through calm seas and, at first, the atmosphere was surprisingly light; but it grew heavier the nearer the boat drew to the orange buoy which rose four feet above the waves and marked the spot and the wreckage beneath.

After a prayer service Molly and the other families cast their wreaths overboard. One young schoolgirl threw over a single red rose on which she had attached a note: 'To our dad (our hero). I will never forget you.'

The captain then started the engines and the resulting ripples carried the flowers towards the buoy, bearing an offering of love and farewell. One widow left the group, stood by the side and carefully began to turn her wedding ring round and round, over the small pleats of skin at her knuckle, down to the end of her finger where it silently slid off. She took it between the forefinger and thumb of her right hand and held it out over the grey water. Then she let go. The wedding band, swallowed by the sea, fell through the first light-filled fathoms and down towards the darkness of the deep until, a moment later, the gold kissed steel and then lay still.

AFTERWORD

I can still remember where I was when I first learned about the Piper Alpha disaster. I was on a week's work experience at the *Evening Times* newspaper in Glasgow, dipping my toes in the inky waters of journalism in which I would contentedly swim for over twenty-five years. The first couple of days had been an eye-opening experience as I attended press conferences and crime scenes, but I hadn't experienced what the wizened reporter I was shadowing described as a 'biggie'. When I arrived early on that Thursday morning the newsroom was a blur of activity as keyboards rattled, reporters dashed around and the subs drew up the front page. I spent most of the day fetching coffees, to which some reporters invariably introduced a swift tot of whisky, but I was intoxicated by the excitement of a big breaking story. I was only sixteen years old but I knew at once this was the world I wished to inhabit. So, in a way, Piper Alpha, a distant disaster, a cause of newsroom commotion, helped inspire my chosen career. It sounds callous, and in a way it is. Journalists are often at their best professionally when others are at their worst physically and emotionally, as they dash to respond to seismic events. Yet I had always been interested in the 'offshore' world. Somewhere buried in an old family photograph album is a picture of me as a child, standing proudly beside an oil rig made of Lego, carefully self-designed and the result of many evenings of creative endeavour. For a brief spell my fantasy career was that of a deep-sea diver, toiling in the dark. Yet it was the type-writer that won out.

I would like to say that after that day in the newsroom I eagerly followed every twist and turn of the aftermath, but I didn't. Like everyone else, except those directly involved, I moved on with my life. I had exams to pass, a future to pursue, and so Piper Alpha, for one day looming large in my life, faded from view. It would be almost eighteen years before I gave the events of 6 July 1988 any serious consideration. I was by then an experienced journalist and a published author in search of a subject for a new book. Looking back I cannot remember what alerted me to the fact that 2008 would be the twentieth anniversary of the Piper Alpha disaster. Yet anniversaries are curious events that momentarily allow the past and present to come together and, albeit briefly, give us a clear view of what came before. Was there a book to be written? One afternoon in the spring of 2006 I wandered upstairs to the *Scotsman*'s cuttings library and asked to see the packets of old newspaper clippings covering the disaster. As a copy boy I'd spent two and half years fetching packets such as these for report-ers, in the days before the internet and digital databases brought everything instantly to hand. I'd developed a fondness for these worn and torn articles, yellowed with age. I carefully unfolded one clipping after another, spreading them out to see what they revealed. It was clearly a hugely complicated story about a distant, mechanized world of which I knew almost nothing. My only real connection to oil platforms was a cousin who worked on the rigs and occasionally brought my father back duty-free pack-ets of King Edward cigars. When researching a new book I'm al-ways looking for a key that can unlock the story, a foothold that will allow me a leg-up to peer inside, or an image that helps me see clearly what the story is actually about. It was in these cuttings that I first met Bob Ballantyne, and after our introduction I knew I had to write the book.

In a newspaper interview shortly after the disaster Ballantyne had talked about how, when he realized that something was wrong, he headed back to his cabin to fetch a book. Unaware of how serious the situation would quickly become, he assumed he and his fellow workers would be taken by helicopter to Norway or Shetland for safety or would spend hours in a supply ship bobbing back to Aberdeen and, if so, who wouldn't wish to be without a book? The novella he collected and stuffed into his survival suit was *Candide* by Voltaire. I couldn't yet understand the complicated world of drilling, condensate pumps, blind flanges or permission-to-work slips, but I could relate to the niggling anxiety of being bookless and the comfort to be found in a worn dog-eared paperback. Then there was Bob's erudite choice of reading material. Who was this man? It was then I realized this project wasn't about mechanics, or hydraulics, or the characteristics of fire, though all would have their place in its pages. No, this was a book about men, ordinary men, and what they will do to survive.

Bertrand Russell said that while researching a new book he would read all he could on a subject until all the pieces coalesced in his mind and then he would simply write what he could see. Piecing together the events of 6 July 1988 was a hugely challenging, emotionally draining, but ultimately, deeply rewarding experience. *Fire in the Night* was researched and written in thirteen months and I can still remember rising at 5 a.m. on 1 January 2007 to begin. Another memory was being emailed the original hardback cover design in the autumn when I had only written a quarter of the manuscript. I printed out an exact size copy and measured the spine then fretted about how many words it would take to fill. The winter as I rushed to finish the book was a dark and difficult time of fevered dreams of smoke, flames and water dark as sin.

The book was published in the summer of 2008 to excellent reviews, most gratifyingly from a novelist who had lost a family member that night. It had also attracted the interest of an Irish film-maker with a view to making a feature film. My agent negotiated a fee for a two-year option. Then we waited, and waited, but in the intervening time the economic crash saw the Celtic tiger strangled with its own tail and the deal was off. It was then that I decided to develop the project myself, not as a film, but as a feature-length documentary. I teamed up with Paul Berriff, the director and cameraman who shot the incredible footage on the night of the disaster, and together with Alan Clements and Michael McAvoy at STV Productions spent almost four years inching towards the big screen.

Paul and I embarked on a grand tour of Britain, tracking down survivors, forgotten film footage, and the audio recordings from the rescue services. When the Wi-Fi went down in our office we worked from the local McDonald's, where the only available seats were in the children's play area. I found myself phoning Occidental, the American owners of Piper Alpha, while sitting on a tiny seat shaped like a toadstool. Occidental thanked us for 'reaching out' to them but declined to assist us in any way. A few months later I received the same brush-off when I travelled to Los Angeles in an attempt to engage with them at their vast, towering offices that loom over Wilshire Boulevard. As we worked on the film we were increasingly aware that Piper Alpha was a forgotten disaster, one which occurred out of sight, far beyond the horizon, 110 miles out to sea, and was now, for the majority of the British population, out of mind.

The documentary focused on a fixed timeline. The disaster began at 10 p.m. with a spark and by midnight the entire platform was destroyed, with the four-storey accommodation block sinking to the bottom of the North Sea. I insisted that it was not a glib 'Hollywood pitch' to describe what took place in that two-hour window as a cross between *The Towering Inferno* and *Titanic*. It was a simple statement of fact. Some of the survivors were approached more in hope than expectation. Joe Meanan, who had given me an interview while I was researching the book, mainly because my card had been marked by a colleague on the *Scotsman* who drank in the pub he owned, had previously refused all television interviews. Yet he felt it was the time to tell his story, and not just his own. Meanan explained that he met, once a year, with two other survivors, Billy Clayton and Dave Lambert, for a few pints in Edinburgh, the mid-point from his own native Aberdeenshire and his colleagues' homes in Newcastle. They too agreed to take part.

Then there was Roy Carey, whose life was saved by a promise he made to his youngest daughter. After diving through a fireball and into the water he surfaced to find himself being cooked alive under a grill of flames. He repeatedly sank under the water for relief then surfaced to the agony of the inferno. Exhausted, he eventually decided it would be less painful to drown but as he sank beneath the surface, for what he hoped would be the final time, he remembered his older daughter's recent wedding and the promise he made to his youngest that she too could have such a day. The emotional jolt gave him the courage to endure, to keep re-surfacing until he was finally able to swim away from the furnace. For some the interviews acted as a form of therapy. John Seabourn, the captain of the *Silver Pit*, said that after we filmed an interview with him, which was sadly not used in the finished film, he enjoyed the best night's sleep since the disaster.

With the research completed we teamed up with Anthony Wonke, the BAFTA-winning and Emmy-nominated director of films such as *Crack House USA*, about a drugs gang brought down by a Federal wiretap, and *The Battle of Marjah*, about a platoon of marines in Afghanistan, and secured financing to make the finished film for a short cinema release prior to broadcast by the BBC on the twenty-fifth anniversary. Anthony took the helm and steered the production with precision, emotional intelligence and artistic brilliance. Brief sequences of reconstruction took the filming to the water tank in Essex used in *The Bourne Supremacy*, while controllable flames were provided by the National Fire College. However the key component to the film's final success was a late discovery of forgotten footage, from a portable VHS video camera, of a tour around Piper Alpha. It was found by our film archivist and was shot by an unknown rigger as he wandered about the platform, joking with friends. Watching the footage for the first time was an eerie, emotional experience as we saw the men laugh and joke, unaware of what would eventually befall them. And there, in the background of the canteen, smiling for the camera, was a young Roy Carey. 'Good lord, that's me,' he exclaimed when we showed it to him for the first time.

Fire in the Night, the documentary, received its world premiere at the Edinburgh International Film Festival in June 2013, where it won the coveted Audience Award for best film. Yet for all the considerable delight of a red-carpet premiere, attended by a number of survivors, it was a smaller screening in Aberdeen the following week that remains the most memorable. For it was in the dark of the Belmond Filmhouse that survivor Roy Thomson finally discovered that he had not killed a man, dispelling a secret guilt he had nursed for almost a quarter of a century. From his cinema seat Thomson watched the screen as Mike Jennings described how he had just been about to jump off Piper Alpha when someone, screaming that their feet were on fire, pushed him from behind and sent him tumbling a hundred feet to the waters below. Roy suddenly realized that he was the man Jennings was talking about. As he later explained: 'I said to my father (immediately after the disaster): "I actually think I killed somebody that night." I've thought that for a long time.' The reason for his misplaced guilt was that when Roy surfaced after his own plunge, the only person nearby was a body, floating face down. 'I just took it that the body I was next to was the guy I pushed off.' Thomson was delighted to discover that the man he believed he had knocked to his death was actually living only a few miles away. They would eventually be reunited, and, as Thomson explained: 'I can finally put my mind at rest.'

There were others for whom guilt, however wrongly imagined, could not be erased. This book begins with Gareth Parry Davies, the diver in the dark depths, and it is only fitting that it ends with him. I never met Gareth, though it was not for want of trying. The opportunity first emerged in the spring of 2009 when I received a phone call at the offices of the *Scotsman* from Terry Parry Davies, Gareth's wife. She had only recently learned about the publication of the book when she googled 'Piper Alpha' to get more information about the disaster for her son and daughter, who were having a hard time with their father's erratic behaviour. Terry always knew it was somehow connected to that night but as Gareth never spoke about it she had no real clear idea about what he had gone through or the heroism he displayed that earned him a bravery award from the Queen. It was only after ordering a copy of the book that she could finally understand the full horror of the night and her husband's place within it.

Gareth Parry Davies was born in Tanzania, but spent his childhood years in England, before leaving Britain to work abroad. He returned to work as a commercial diver in the North Sea in 1987, a year before the disaster, and was haunted by that night ever after. Although he did his best to dispel any dark thoughts with constant motion, pursuing a string of adrenaline-fuelled jobs such as diving for diamonds in Angola and Zaire, he was unable to outrun them, no matter how far he travelled. He felt guilty for simply surviving when 167 men had perished, and on his shoulders lay another cross, one of his own construction as no one would have foisted it upon him, but still one whose weight he felt keenly. At the Cullen Inquiry into the disaster, Gareth learned that the sprinkler system was inactive that night. The reason it was inactive was a company policy which dictated that it be switched off whenever there was a diver in the water. He believed that if he had not been in the water, more men might have survived.

When it was time to film the documentary I contacted Terry and asked if Gareth would meet with me. He said he would but didn't show up at the restaurant, leaving Terry and me to dine alone. A second meeting, for coffee, was equally one-sided. We spoke on the phone a couple of times, mainly about books. Gareth was a fan of Hunter S. Thompson and when the gonzo journalist's collected letters were published I sent him a copy as a present. He phoned up to ask, perhaps only half in jest, if this was 'a fucking bribe?' I'm told he felt my book *Fire in the Night* was true and accurate, but in the end he wanted no part in the documentary.

In 2012 Gareth moved from Colchester to Limerick in Ireland, where he set up his own cleaning business, and when he last spoke to Terry, from whom he was separated but remained on good terms, she said he seemed upbeat and happy. He was staying in an apartment near the River Shannon, in whose waters his body was found on 10 June 2015. He was fifty-nine years old and both Terry and his close friend Adrian Cassavella believe he was, ultimately, the last victim of Piper Alpha. As Adrian said: 'The ghost of Piper Alpha never really left him, and although it was a subject he avoided, the impact became more and more evident.' As Gareth wrote in a short note he left for Terry: 'If I could have found a way to replace the darkness with light I would have.'

I didn't attend the funeral in Limerick, but sent a short note, which ended: 'As his friends and family gather today to mourn his death, all I can do is remind them in their grief and sorrow and, yes, anger, that men are alive today because of the bravery of Gareth Parry Davies. Few people leave such a legacy.'

But what was the legacy of Piper Alpha for the oil and gas industry? There can be little doubt that the North Sea, and many offshore installations around the world, are safer today as a consequence of the changes made by the industry in the immediate aftermath. Yet a platform, in the middle of the North Sea, funnelling oil and gas is no place for complacency, and these remain difficult times for the oil industry. In the last four years the collapse of the price of oil has led to a severe downturn and the loss of 160,000 jobs in the maritime oil and gas industry. The North Sea basin is now characterized as 'ultra-mature' and has been described in *The Economist* as a 'maritime used-car lot with second-hand rigs changing hands at low prices and old bangers being decommissioned or turned to scrap.' Although still afloat, many Scots in the industry insist it would be a foolhardy man who pipes a final lament for the North Sea industry, but my concern is that in straitened times costs will be cut and once again safety will take a secondary place to profit. My one hope is that both the book and documentary of *Fire in the Night*: *The Piper Alpha Disaster* continue to act as a reminder of the terrible price that could again be paid by the men and now women who work far beyond the horizon to keep our cars moving and homes warm. Thirty years on Piper Alpha is now modern history. Let us hope it never again becomes a current event.

Acknowledgements

The decision to write an account of 6 July 1988 and the destruction of Piper Alpha was not lightly taken, and the worry of reopening old wounds among survivors and the bereaved was a constant companion. However, I have been encouraged by the number of people who welcomed an accessible account of the world's worst offshore oil disaster, especially at a time when a new generation of young Scots is entirely ignorant of the human price paid beyond the horizon for oil and gas. I would like to especially thank Molly Pearston, who lost her son Robert, for her support and encouragement, and Janis and Troy Busse, who spoke about the loss of their brother, Carl Busse. A number of survivors assisted me in a variety of ways and I would like to thank William Barron, Michael Jennings, Joseph Meanen, Edward Punchard, Steven Rae, Geoff Bollands and David Kinrade, as well as Pat Ballantyne, the widow of Bob Ballantyne, and James McNeill.

The oil industry is a complicated machine and a number of people assisted me in my understanding of how it works. At the Offshore Industry Liaison Committee (OILC) I would like to thank Jake Molloy, the general secretary; Lorna Robertson, head of administration; and Ronnie McDonald. At Oil & Gas UK I would like to thank Britta Hallbauer, Chris Allen and Sally Fraser, as well as Jenny Costelloe and Brian O'Neill at TOTAL E&P UK and the author Bill Mackie. Any errors that slipped through are entirely my responsibility. Among those journalists who assisted me with their recollections of the event were my colleague on the *Scotsman*, Frank Urquhart; Jane Franchi at the BBC;

ACKNOWLEDGEMENTS

Donald John MacDonald at STV North, who generously allowed me access to the archives of Grampian News; and Paul Berriff, who organized a copy of the footage he filmed on the night of the disaster. I would also like to thank Steve Snyder, managing editor of the *Navasota Examiner* in Texas.

My understanding of the response by Grampian Police was assisted by Ian Gordon and Alastair Ritchie, who gave generously of their time and experience. I would also like to thank Doctor Graham Page at Aberdeen Royal Infirmary, Alan Reid, Dr Stephen Hearns, David Burnside and Frank Lafevre. A champion and early supporter of this book was Professor David Alexander, director of the Aberdeen Centre for Trauma Research. I would also like to thank Dr David Drysdale, Dr Terry Brotherstone, David Tumelty, Sir Ranulph Fiennes and the artist Sue Jane Taylor. At the *Scotsman* I would like to thank Mike Gilson, my editor, Ian Stewart, Paul Riddell, Frank O'Donnell and the denizens of the Glasgow office, Craig Brown, Martyn Mclaughlin, Alastair Dalton, Emma Cowing and Lyndsay Moss, as well as Alex Hewitt on the picture desk. Thanks must also go to David Wylie, Frank Deasy and Allan Clark, to my agent, Eddie Bell, and all the team at Bell-Lomax, while at Macmillan I must thank my editor Richard Milner, Lorraine Baxter, and all the team who ushered this book into print.

I am particularly indebted to Vernon Baxter, who assisted me in the early stages of the research. He is a young journalist who is going places – Dubai, in fact.

Last in the acknowledgements but first in my life, my most sincere thanks and love must go to my wife, Lori, who gave me greater assistance than she will ever know.

Stephen McGinty,
March 2008

Index

INDEX

INDEX

extracts reading groups
competitions books new events
discounts extracts extracts discounts
competitions extracts reading groups extracts
books new reading groups extracts discounts
events books extracts events
extracts reading groups
new title reading groups
interviews reading groups books
reading books events extracts events new
discounts books interviews books
new books events events
events new interviews new books extracts
discounts extracts discounts books

www.panmacmillan.com

extracts events reading groups
competitions books extracts new books